I0053265

CT Scanning of Carbonate Reservoirs

The Computed Tomography (CT) scanning of carbonate reservoirs is a non-destructive method used to obtain valuable information from reservoir rocks. This concise book covers all aspects of CT image analysis and their interpretation. It is focused on the CT scanned images and the data gathered from various carbonate reservoirs with different ages, from Paleozoic to Tertiary, and with different textures, degrees of heterogeneity, facies types, diagenetic impacts, fossil contents, sedimentary structures, reservoir quality, and fracturing. Numerous high-resolution images illustrate various aspects of carbonate rock properties and are suitable for analysis by both professionals and students.

FEATURES

- Provides the first specialized book about core CT scanned image descriptions and explanations
- Contains supplementary data from various sources, such as image logs, cores, and thin section photos and petrophysical measurements
- Includes original CT images from different reservoirs
- Covers both research and industrial aspects of core CT scanning and is useful for comparing CT scanned images
- Uses original images and reservoir data gathered by the author

This book can be used by professionals working in the oil and gas industry, researchers and academics studying petroleum-related disciplines, and students taking courses in petroleum geology and reservoir engineering. It can also be used in a range of related sciences, such as sedimentology, soil sciences, paleontology, petrology, engineering geology, study of porous materials, and hydrogeology.

CT Scanning of Carbonate Reservoirs

A Color Atlas

Vahid Tavakoli

CRC Press
Taylor & Francis Group
Boca Raton London New York

CRC Press is an imprint of the
Taylor & Francis Group, an **informa** business

First edition published 2024
by CRC Press
2385 NW Executive Center Drive, Suite 320, Boca Raton FL 33431

and by CRC Press
4 Park Square, Milton Park, Abingdon, Oxon, OX14 4RN

CRC Press is an imprint of Taylor & Francis Group, LLC

© 2024 Vahid Tavakoli

Reasonable efforts have been made to publish reliable data and information, but the author and publisher cannot assume responsibility for the validity of all materials or the consequences of their use. The author and publishers have attempted to trace the copyright holders of all material reproduced in this publication and apologize to copyright holders if permission to publish in this form has not been obtained. If any copyright material has not been acknowledged please write and let us know so we may rectify in any future reprint.

Except as permitted under U.S. Copyright Law, no part of this book may be reprinted, reproduced, transmitted, or utilized in any form by any electronic, mechanical, or other means, now known or hereafter invented, including photocopying, microfilming, and recording, or in any information storage or retrieval system, without written permission from the publishers.

For permission to photocopy or use material electronically from this work, access www.copyright.com or contact the Copyright Clearance Center, Inc. (CCC), 222 Rosewood Drive, Danvers, MA 01923, 978-750-8400. For works that are not available on CCC please contact mpkbookspermissions@tandf.co.uk

Trademark notice: Product or corporate names may be trademarks or registered trademarks and are used only for identification and explanation without intent to infringe.

ISBN: 978-1-032-52140-4 (hbk)
ISBN: 978-1-032-52141-1 (pbk)
ISBN: 978-1-003-40541-2 (ebk)

DOI: 10.1201/9781003405412

Typeset in Palatino
by SPi Technologies India Pvt Ltd (Straive)

Contents

Preface

X-ray computed tomography (CT scanning) uses X-ray beams to detect internal structure of the objects. This technique is frequently used in core analysis of hydrocarbon reservoirs. In addition to whole cores, plugs are also scanned to trace any special feature such as fractures or stylolites. Resulted images are also used in digital petrophysics, which employs the CT images to understand the geological and petrophysical properties of hydrocarbon bearing formations. As this is a non-destructive, robust method to reveal internal structure of the rocks, its application has significantly increased in recent years. A considerable number of reservoir properties can be understood by using this method.

Despite all of these advantages, few studies have been published about CT scan image analysis and interpretation. How 3D geological features are traced in 2D CT slices? How porosity and various pore types are seen on these images? What enhancement techniques are applied to these images and how the results are interpreted? How CT scan images are compared with the other sources of information from carbonate reservoirs? This book tries to answer these questions. It starts with an introduction about principles and applications of core CT scanning. In this regard, the scientific background of the technique is considered. Some examples are presented about the application of this method in petroleum geology and reservoir engineering. Also, the introduction of the first chapter explains how this book can be used. It explains the legends and signs which have been used in the figures.

In Chapters 2 and 3, each page is composed of an original CT photo, a traced photo (the same image which is marked with arrows and lines), FMI, whole core, close-up and thin section photos. Sample information is also provided, which includes age, main lithology and porosity of the sample, as well as the type of light in thin section photography. Chapter 4 considers various methods to enhance the image visualization for better interpretation or analysis. There are two images in these pages. Routinely, the upper image is the original CT slice while the lower one has been filtered or segmented. This is different in some cases, when viewing the original photo was not necessary. The first image also presents enhancement technique in these cases.

The book covers both academic and industrial aspects of the CT image analyses and interpretations. The main application of the book is in petroleum geology and reservoir engineering, but it can also be used in sedimentology, soil sciences, paleontology, petrology, engineering geology, study of porous materials, hydrogeology and other related sciences. Therefore, this book would be useful for any student, researcher, industry professional or anyone who wants to know the internal structure of a material with a non-destructive technique.

Preparation of this book was made possible through my experiences in core analysis projects of various petroleum fields. CT scan images and data from various carbonate reservoirs with different ages are considered from Paleozoic to Tertiary with different textures, degree of heterogeneity, facies types, diagenetic impacts, fossils contents, sedimentary structures, reservoir quality and fracturing. Such as always, I am thankful to Dr. M. Naderi-Khujin for her endless support in my researches. She studied the provisional version of the book and gave me brilliant ideas about it. I want to thank all of my colleagues and students in University of Tehran which are developing the earth sciences.

Please let me know any idea, critique or suggestion which you probably have about the book.

Vahid Tavakoli
University of Tehran
vtavakoli@ut.ac.ir
April 2023

Author

Vahid Tavakoli, PhD, is an associate professor at the University of Tehran, Iran. In 2011, he earned a PhD in petroleum geology at the University of Tehran. Since 2003 he has worked in the field of reservoir characterization of the Iranian hydrocarbon formations using cores and wireline logs data. He is the author of two books, *Geological Core Analysis: Application to Reservoir Characterization* and *Carbonate Reservoir Heterogeneity: Overcoming the Challenges*, published in 2018 and 2020, respectively. In 2018, Dr. Tavakoli was honored with the Young Geologist Award by the Iranian Academy of Sciences. He has published over 50 scientific papers in high-ranked international journals.

1

Principles and Applications of Core CT Scanning

1.1 Introduction

X-ray computed tomography (CT scanning) prepares cross-sectional images from an object. Computers are employed to combine X-ray measurements from various angles and present the results with two- or three-dimensional images. The internal objects of the matter can be observed with this method. At first, it was developed for medical purposes in the early 1970s (Hounsfield, 1972, 1973). Then, its application in petroleum geology was soon recognized (Vinegar, 1986; Vinegar and Wellington, 1986). The technique can be used for both quantitative and qualitative analyses of internal features of the rocks. Materials with different densities (electron density and photoelectric factor) can be distinguished with this method. Highest contrast exists between various material phases. Therefore, it plays an important role in the study of porosity and pore network of the rocks. Regarding this application, CT scanning has wide use in soil sciences, hydrogeology and specially petroleum geology and formation evaluation. Comparison of the results with more traditional methods, such as thin section studies, X-ray diffraction (XRD), scanning electron microscopy (SEM) and routine core analysis (RCAL) should be made in many cases. As X-rays have no interaction with the samples (at least in geological studies), it can be used for monitoring fluid movement (Chen et al., 2001; Yasuda et al., 2018; Wang et al., 2023) and deformation (Zhao et al., 2014; Hu et al., 2022). Both ambient and reservoir conditions (temperature and pressure) can be used in these tests. Therefore, it is possible to understand the rock and fluid interactions in dynamic situations. Also, unique and valuable samples can be analyzed with this method without any destruction. In this regard, various studies have been carried out on fossils (e.g., Brombacher et al., 2022; Yu et al., 2022; Bosch et al., 2023) and meteorites (e.g., Néri et al., 2020).

Recently, CT scanning has acquired a great importance in reservoir studies (Tavakoli, 2018a; Bera and Shah, 2021; Haagsma et al., 2021; Saraf and

DOI: 10.1201/9781003405412-1

Bera, 2021; Chawshin et al., 2022; Li et al., 2022; Eltom and Goldstein, 2023; Kaveh-Ahangar et al., 2023; Mees et al., 2003; Tan et al., 2023). It is a rapid, low-cost, non-destructive and three-dimensional technique which could be used in many aspects of formation evaluation. Despite such importance, relatively few works have been focused on illustrating the reservoir rock properties in CT slices.

This book shows various geological and petrophysical characteristics of the carbonate reservoir rocks in CT slices of the cores. Both whole cores and close-up images are presented to show various aspects of the technique. Results are compared with image logs, mainly Fullbore Formation MicroImager (FMI), thin section data and photos and petrophysical properties. Examples of image enhancements, segmentation, binarization, filtering, rasterizing and clustering are also provided.

The first pages are about the principles and applications of CT scan data and images in carbonate reservoir studies. They illustrate the application of this non-destructive method in various geological and petrophysical studies with examples from different parts of the world. Then, sample selection using CT images is discussed. The book continues with consideration of geological properties including facies and diagenesis in CT images. Rock textures from mudstone to wackestone, packstone and grainstone are considered. Various allochems in different textures are presented. Examples of microbial facies (stromatolites and thrombolites) are also included to show the effects of these textures on CT images of the carbonates. Thin section and core photos are presented along with CT images to clearly show the discussed properties of rocks. Various diagenetic processes such as dolomitization, dissolution, cementation, fracturing and replacement are illustrated. As fracture study is one of the most important applications of CT scanning, various fracture types are considered on CT sections. Fractures are also evident on image logs and so they are also used to support the image interpretations. Pore types strongly affect the petrophysical properties of carbonate reservoir rocks, and therefore, the book discusses how pore types could be distinguished in CT images. Petrophysical properties derived from the RCAL are provided to support the interpretations.

The effects of different techniques such as image segmentation, clustering, binarization, filtering and rasterizing are also considered. Examples are presented which enable the reader to consider the effect of various image analysis techniques on CT slices. Results are compared with geological and petrophysical data.

Legends are also provided (Figure 1.1), which explain the various parts of the pages. Generally, pages have two different formats. Two transversal CT images are seen in planar structures and sedimentological (facies and diagenetic) pages (Chapters 2 and 3). The upper photo is original output. Just brightness and contrast have been changed in this photo for better view. The second one is the same as the first, which has been traced to show the properties of interest. In the right column, both whole core photo and FMI

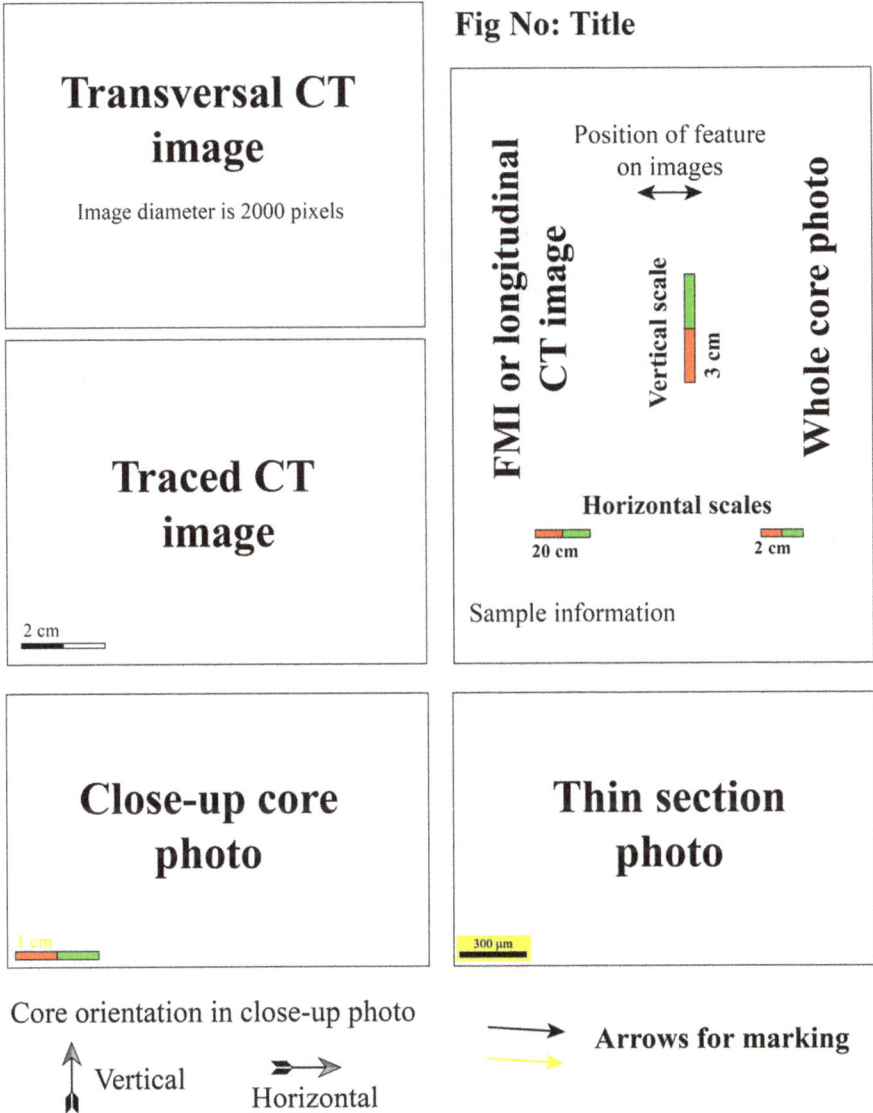

FIGURE 1.1
Legend of the pages. The number beside each horizontal bar is the size of one segment. The close-up scale defines the size of the width of the core.

image log can be seen. Geologists use these two sources for the study of planar structures, and therefore this is important to show these structures on both of them. Close-up core photo and image of relevant thin section are presented at the lower part of the page to completely show the nature of the structure or the facies and texture of the studied rock. Sample information including, age, main lithology, type of light in thin section photography and porosity are also written in the right part of each page. A description is provided at the bottom of page, which explains the property of interest in the CT image. It also provides a brief explanation about the subject.

The format of image segmentation and enhancement section (Chapter 4) is different. The CT images for this part are selected from previous images, which their whole core, close-up, FMI and thin section images have been presented before. Routinely, this part has two images. The upper one is the original image and the lower has been enhanced or segmented to show the application of various image analysis techniques to CT slices. Description of the technique and sample properties are also presented. In most cases, the explanation is supported by petrophysical data from core analysis or thin section data from petrographical studies.

The value of porosity in all cases comes from the nearest point, which has maximum distance of 15 cm. Porosity has been measured on plug samples using Boyle's law. Air has been employed as the test fluid.

1.2 Principles of CT Scanning

In geology, CT scanning is performed to obtain information about the properties and position of materials occupying inside a rock. Images are produced based on the attenuation (reduction in intensity) of X-rays which are recorded by a series of detectors. Therefore, there are three basic components in each CT scanner including an X-ray source, a rotation system in 3D space and a detector. As 3D rotation needs more physical space, the tube-like source routinely has a linear motion around the subject. Hence, the subject is completely scanned. Clinical CT scanners are usually used for core samples (Figure 1.2), but there are also core scanners which have been designed for CT scanning of rock samples.

Cores are scanned before opening. Therefore, core sleeve and fasteners are also present at the time of scanning. Both sequential and spiral scans are used. In sequential scanning, the core sample is moved and a slice is prepared in predesigned positions. In spiral scanning, the X-ray tube rotates around the sample and continuous data is prepared. The parallel cross-sections can be reconstructed to build and display a 3D shape of the sample. This is achieved by computer programs and special mathematical formulas, which are beyond the scope of this book.

X-ray photons mainly interact with electrons. As the electron density in materials is almost the same as their bulk density, the attenuation is different for various materials and strongly depends on their density and mean atomic number. Compton scattering and photoelectric absorption mainly affect the X-ray attenuation and resulted image. The attenuation of high-energy X-rays

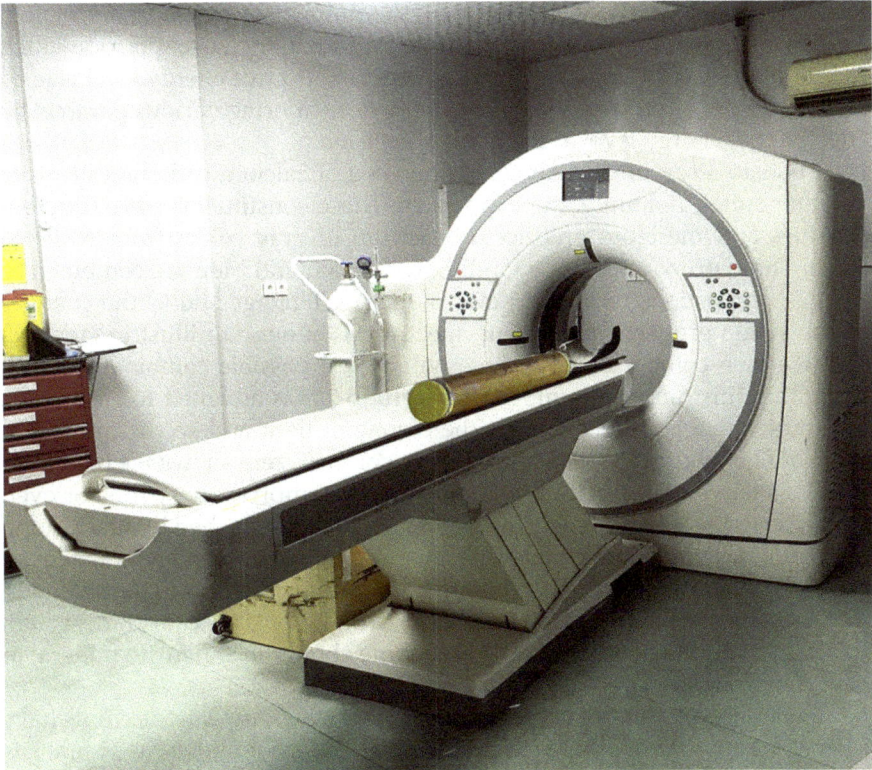

FIGURE 1.2
A clinical CT scanner is routinely used for scanning of core samples. Cores are scanned prior to opening (courtesy of Nadia Naderi).

is mainly determined by Compton scattering while photoelectric absorption is more important at lower energy levels.

Volume elements or voxels are parts of the sample with equal, square-shaped cross sections. Suppose that the geologist is interested in a cross-sectional slice of a core body that is, say, 10 cm thick. The sample can be subdivided into five, 2 cm long blocks. These blocks are called voxels. It is worth mentioning that the term is also defined in other sciences. For example, in a reservoir numerical model, a voxel is a cube which forms the basic volume of the model. The X-ray is linearly attenuated in a voxel depending on the properties of materials. Based on the quality of the sources and detectors, as well as the desired resolution, the size of the voxels is determined. Knowing the size of the sample and voxels, the number of voxels in the sample is calculated. In resulted cross section, the X-ray attenuation in each voxel is depicted as a 2D square. The square has uniform grayness which is proportional to the attenuation in the voxel. These squares are called picture elements or pixels. Each CT image is composed of some pixels and the image resolution is determined by the number of pixels in a specified area. The beam attenuation may differ according to the voxel position in the rock and the energy of the beam. Therefore, an effective energy level is used for calculations. The number is achieved by considering various parameters affecting the energy level at the point of incision.

Carbonate reservoirs are mainly composed of calcium carbonates (calcite and aragonite), dolomite and evaporites. These constituents have different densities and therefore produce a spectrum of gray colors from white to black. Three primary colors including red, green and blue are combined in various proportions to obtain new colors. A CT image shows the changes in one single property and so one spectrum is enough to illustrate the picture. As each color is one bit, there will be 256 possible combinations (2^8). Regarding this range, a number from zero to 255 is assigned to each pixel. Computers display the colors by the degree of light and so the maximum number (255) indicates the white color. In contrast, zero shows the black and various shades of grays form a CT image. The empty pore spaces, or even fluids, have the lowest density and so they are distinguished by black color. Clay minerals and organic materials have also low densities. Mineral density increases in quartz, calcite and dolomite, as routine minerals in the sedimentary rocks, respectively (Figure 1.3). Instead, high-density heavy minerals are specified by light spots. This is also the same about the anhydrite mineral in a carbonate rock.

Many interpretations of CT images are qualitative. Anyway, attempts have been made to quantify these interpretations. The most widely used index is CT number. It is defined as:

$$CT\,number = \left(\left(\mu_{material} - \mu_{water} \right) / \mu_{water} \right) \times 1000$$

where μ is the attenuation coefficient for the energy beam and CT number is expressed in Hounsfield Unit (HU). In this scale, the CT number of vacuum is -1000, water is zero and a dense bone is 1000. The attenuation coefficient indicates how easily the beam can penetrate into the desired material. In fact, it shows the weakening of X-ray beam after passing through a substance. Energy-dependence of attenuation is unique for each material and these are different from water. Therefore, the CT number is different in different energy levels. A standard energy level should be defined to obtain reproducible results. Normalizing attenuation coefficient between maximum and minimum has been used in some researches (e.g., Hsieh, 2012). Different carbonate reservoir features can then be identified by this technique. It is worth mentioning that unusual heavy minerals (such as pyrite) should be excluded before such normalization. Many attempts have been made to quantify CT results in geological studies (e.g., Ketcham, 2019; Yan et al., 2021), but there is still considerable discussion about a global standard for geological materials.

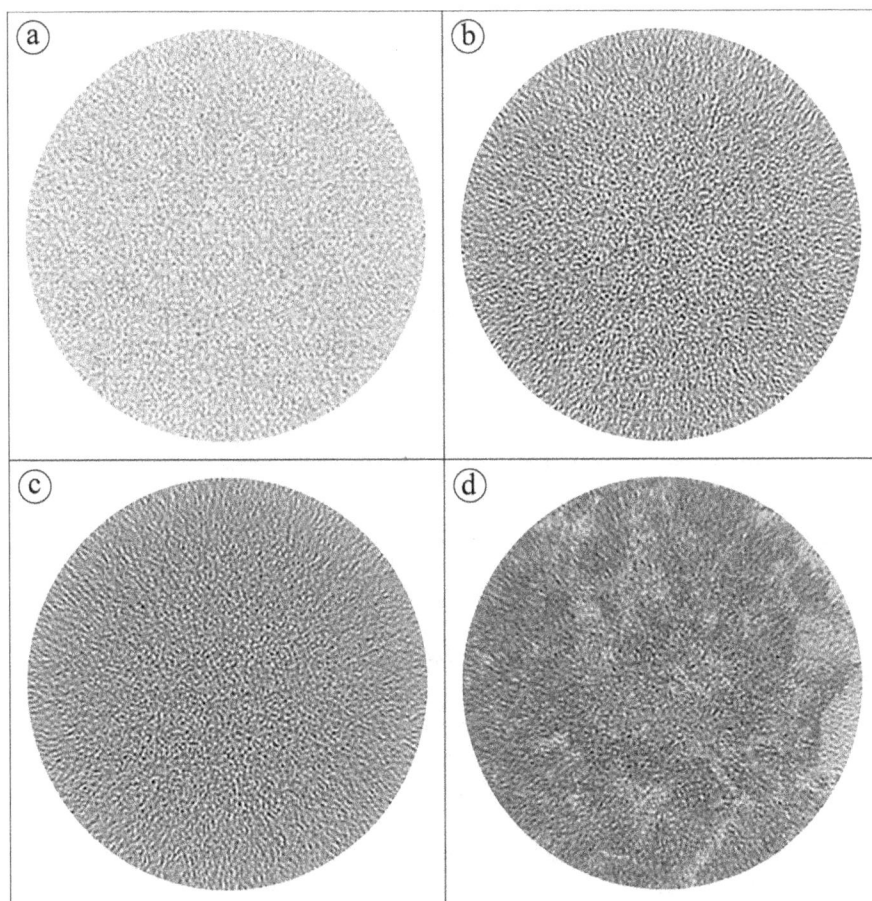

FIGURE 1.3
Comparison of four lithologies of carbonate reservoirs. Anhydrite (a) is light gray while dolomite (b) is darker. The darker colors are more obvious in lime (c). The difference is more obvious in a sample with all lithologies (d) (lime: 35%, dolomite: 50% and anhydrite: 15%). All images have been prepared with the same conditions. Porosity is less than 3% in all images.

1.3 Applications in Petroleum Geosciences

At first, CT scan images were used for distinguishing fractures and their characteristics in the cores (e.g., Honarpour et al., 1986; Bergosh and Lord, 1988; Chen et al., 2001). Core opening, layout and marking may change the fracture properties and therefore it is important to consider them before opening the cores. Open fractures contain formation fluids and therefore have less density than the surrounding materials. Filled fractures in carbonates routinely contain denser cements in comparison to rock matrix because evaporites, especially anhydrite, have more density than calcite and dolomite. Advances in CT scan techniques caused more attention to this method. Digital petrophysics, the science of formation evaluation based on the digital photos, mainly uses CT images (Liu et al., 2015; Saxena et al., 2019; Miarelli and Della Torre, 2021). Today, many reservoir characteristics are derived from core CT-scan data. These include, but are not limited to, facies analysis, diagenetic properties, porosity, pore types, permeability, water saturation, formation resistivity factor, formation resistivity index, mechanical and acoustic characteristics, formation damage and enhanced oil recovery. Interpretation of these properties from images and CT data requires a combination of experience, image enhancement techniques and comparison with other types of reservoir data.

Unconventional energy resources (such as tar sands, shale oil and gas and tight reservoirs) are now one of the primary targets for oil and gas exploration and production. Routine analyses of hydrocarbon bearing reservoir rocks are unable to accurately address the properties of these reservoirs. This is due to their complex pore systems and fluid flow behavior (Le and Murad, 2018; Nazari et al., 2019; Al-Rbeawi and Al-Kaabi, 2020; Bera and Shah, 2021). As CT scan technique is able to show the objects inside the rocks, it has attracted the interest of researchers to analyze these unconventional reservoirs.

Both static and dynamic analyses have been performed using CT images. Basso et al. (2021) used CT images of a lacustrine carbonate reservoir from the Santos Basin to recognize the facies characteristics and petrophysical properties of the samples. The attenuation coefficients of the various parts of the rocks were considered. They concluded that CT technique can be used for distinguishing minerals of the samples. Quartz has less X-ray attenuation compared to calcite and dolomite and therefore can be distinguished with darker gray values. Images were also used for identifying heavy minerals, which are characterized by bright spots in the images. They stated that these minerals are associated with some facies and pore types and therefore can be used for reservoir evaluation. Nunes et al. (2022) combined computerized tomography with thin section studies and defined tomofacies in Morro do Chaves Formation, Brazil. Tomofacies were compared with rock types, wireline log data and RCAL results. A good correlation was observed between these data. Then, boundary values were achieved for distinguishing reservoir and non-reservoir parts. The studied formation is composed of carbonate, siliciclastic and hybrid facies. They proved that a multidisciplinary approach is useful to characterize and evaluate the various factors that control reservoir properties of a heterogeneous sequence. It is also useful for determining producing intervals of hydrocarbon bearing formations. Wang et al. (2022) tried to construct water saturation models in fractured carbonates using CT scan images. They made a fractured digital rock sample using CT scan data and then the oil-water distribution was simulated in this model. At last, a saturation model was proposed and compared with real measurements. They concluded that results are more consistent with the measurements and so the saturation model is more applicable in similar formations. Duan et al. (2023) used CT scan data for distinguishing pore type distribution and interpreting acoustic velocity dispersion in carbonate reservoir rocks. They concluded that pore network structure has great impact on the dispersion of sonic wave in carbonates.

Machine learning techniques are frequently used in petroleum industry. They are also useful for data analyses of CT images. These methods have been used for porosity and permeability prediction, image segmentation, fossils' identification, upscaling and downscaling of data (e.g., Menke et al., 2021; Haagsma et al., 2021; Zhao et al., 2022; Liu et al., 2023).

1.4 Sample Selection

Routinely four plugs are prepared from each meter of core for conventional and geological core analyses (Tavakoli, 2018b). It is not possible to perform special core analysis tests on all of these samples. These tests are expensive and time consuming and therefore, sample selection is mandatory. Also, some samples should be selected for whole core analysis, which measures petrophysical properties on whole cores. This is also the case about the rock mechanical tests on the cores. The selection strongly depends on the degree of heterogeneity of the formation. Samples are grouped using various methods such as Winland R35 (Kolodizie, 1980), flow zone indicator (FZI) and reservoir quality index (RQI) (Amaefule et al., 1993), Lucia rock fabric number (RFN) (Lucia and Conti 1987; Lucia, 1995), pore facies classification (e.g., Tavakoli et al., 2011) and geological rock typing (Tavakoli, 2018b). Some samples are selected from each group. These samples must be good representatives of all samples. Therefore, they should be carefully checked before any experiment. This check is routinely performed by CT scan of the plugs or cores. More slices are prepared in these cases with closer spacing. Presence of some features such as fractures, stylolites, solution seams, connected vuggy pores and barrier layers (shale, anhydrite) will result in wrong measurements. In fact, such samples are not good candidates for the entire formation, or at least the rock matrix properties. Figure 1.4 shows seven transversal slices of a core as well as its longitudinal section (Figure 1.4a). This sample had been selected for whole core analysis. It is clearly seen that at least four fractures are present in the sample. The first transversal slice (Figure 1.4b) shows two open fractures. They continue until section d (Figure 1.4c and d). Vuggy and moldic pores are frequently seen in e and f sections. There are two fractures in slice h. Parts of the left-side fracture are filled with anhydrite while the right-side one is a closed fracture. Any fluid can easily pass through these fractures and so the result will not show the real properties of the whole rock. It is worth mentioning that fractures are also part of the reservoir and so they should be considered in combination with other data.

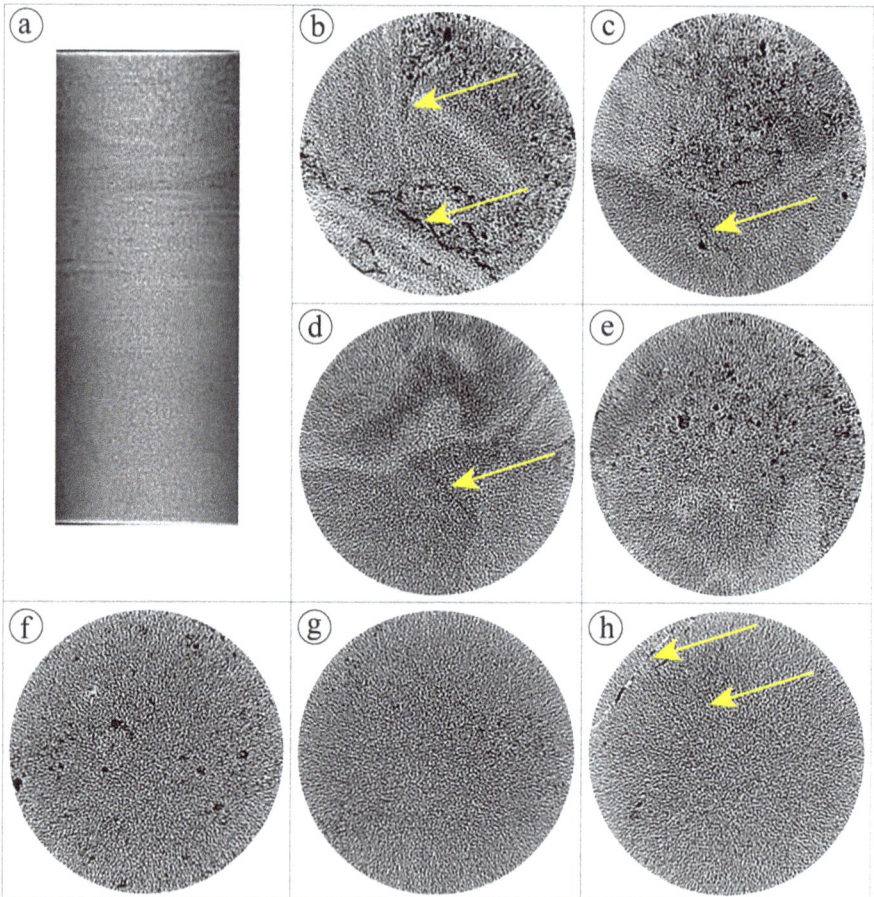

FIGURE 1.4
A whole core has been scanned for special core analysis. Results show fractures in the core.

References

Al-Rbeawi S., Al-Kaabi M.H.H. A hybrid model for the combined impact of non-Darcy flow, stimulated matrix permeability, and anomalous diffusion flow in the unconventional reservoirs (2020) *Upstream Oil and Gas Technology*, 5, art. no. 100020.

Amaefule J.O., Altunbay M., Tiab D., Kersey D.G., Keelan D.K. Enhanced Reservoir De-scription: Using core and log data to identify hydraulic (flow) units and predict permeability in uncored Intervals/Wells (1993) In: *68th Annual Tech Conf and Exhibit*. Houston, TX Paper SPE26435.

Basso M., Belila A.M.P., Chinelatto G.F., Souza J.P.P., Vidal A.C. Sedimentology and petrophysical analysis of pre-salt lacustrine carbonate reservoir from the Santos Basin, southeast Brazil (2021) *International Journal of Earth Sciences*, 110 (7), pp. 2573–2595.

Bera A., Shah S. A review on modern imaging techniques for characterization of nanoporous unconventional reservoirs: Challenges and prospects (2021) *Marine and Petroleum Geology*, 133, art. no. 105287.

Bergosh J.L., Lord G.D. New developments in the analysis of cores from naturally fractured reservoirs (1988) *Society of Petroleum Engineers of AIME*, (Paper) SPE, pp. 57–64.

Bosch M.D., Buck L.T., Strauss A. Perforations in Columbellidae shells: Using 3D models to differentiate anthropogenic piercing from natural perforations (2023) *Journal of Archaeological Science: Reports*, 49, art. no. 103937.

Brombacher A., Searle-Barnes A., Zhang W., Ezard T.H.G. Analysing planktonic foraminiferal growth in three dimensions with foram3D: An R package for automated trait measurements from CT scans (2022) *Journal of Micropalaeontology*, 41 (2), pp. 149–164.

Chawshin K., Berg C.F., Varagnolo D., Lopez O. Automated porosity estimation using CT-scans of extracted core data (2022) *Computational Geosciences*, 26 (3), pp. 595–612.

Chen H.L., Lucas L.R., Nogaret L.A.D., Yang H.D., Kenyan D.E. Laboratory monitoring of surfactant imbibition with computerized tomography (2001) *SPE Reservoir Evaluation and Engineering*, 4 (1), p. 16.

Duan X., Zheng H., Liu X., Liang L., Xiong J. Study on acoustic velocity dispersion of carbonate rock and extrapolation of the velocity (2023) *Acta Geophysica*, 71 (2), pp. 723–733.

Eltom H.A., Goldstein R.H. Scale dependence of petrophysical measurements in reservoirs with Thalassinoides: Insights from CT scans (2023) *Marine and Petroleum Geology*, 148, art. no. 106036.

Haagsma A., Scharenberg M., Keister L., Schuetter J., Gupta N. Secondary porosity prediction in complex carbonate reefs using 3D CT scan image analysis and machine learning (2021) *Journal of Petroleum Science and Engineering*, 207, art. no. 109087.

Honarpour M.M., McGee K.R., Crocker M.E., Maerefat N.L., Sharma B. Detailed core description of a dolomite sample from the upper madison limestone group (1986) *Society of Petroleum Engineers of AIME, (Paper) SPE*, pp. 501–512.

Hounsfield G.N. A method of and apparatus for examination of a body by radiation such as X- or gamma-radiation (1972) British Patent No 1.283.915, London.

Hounsfield G.N. Computerized transverse axial scanning (tomography). Part 1: Description of system (1973) *British Journal of Radiology*, 46, pp. 1016–1022.

Hsieh, C.H. Procedure and analysis of mineral samples using high resolution X-ray micro tomography (2012) The University of Utah, Master thesis.

Hu S., Yu Y., Cheng W., Zheng L., Lv T. Study on the effect of injection rate on the fracture initiation and extension in cement-sand slurry specimens (2022) *Bulletin of Engineering Geology and the Environment*, 81 (4), art. no. 150.

Kaveh-Ahangar S., Nozaem R., Tavakoli V. The effects of planar structures on reservoir quality of Triassic Kangan formation in the central Persian Gulf, an integrated approach (2023) *Journal of African Earth Sciences*, 197, art. no. 104764.

Ketcham R.A. Resolution-invariant measurements of small objects in polychromatic CT data (2019) *Proceedings of SPIE - The International Society for Optical Engineering*, 11113, art. no. 111130B

Kolodizie S.J. Analysis of pore throat size and use of the Waxman-Smits equation to determine OOIP in Spindle Field, Colorado (1980) SPE paper 9382 presented at the *1980 SPE Annual Technical Conference and Exhibition*, Dallas, Texas.

Le T.D., Murad M.A. A new multiscale model for flow and transport in unconventional shale oil reservoirs (2018) *Applied Mathematical Modelling*, 64, pp. 453–479.

Li X., Ji H., Chen L., Li M., Xu K., Jiang X., Zhang Z., Zhang Z., Guo X. Hydraulic fractures evaluation of the glutenite and the effects of gravel heterogeneity based on cores (2022) *International Journal of Rock Mechanics and Mining Sciences*, 160, art. no. 105264.

Liu J.-J., Song R., Cui M.-M. Improvement of predictions of petrophysical transport behavior using three-dimensional finite volume element model with micro-CT images (2015) *Journal of Hydrodynamics*, 27 (2), pp. 234–241.

Liu M., Ahmad R., Cai W., Mukerji T. Hierarchical homogenization with deep-learning-based surrogate model for rapid estimation of effective permeability from digital rocks (2023) *Journal of Geophysical Research: Solid Earth*, 128 (2), art. no. e2022JB025378.

Lucia F.J. Rock-fabric/petrophysical classification of carbonate pore space for reservoir characterization (1995) *AAPG Bulletin*, 79(9), 1275–1300.

Lucia F.J., Conti R.D. Rock fabric, permeability, and log relationships in an upward-shoaling, vuggy carbonate sequence (1987) *Bureau of Economic Geology, Geological Circular 87–5*, 22. doi.org/10.23867/gc8705D.

Mees F., Swennen R., Van Geet M., Jacobs P. (eds) Applications of X-ray computed tomography in the Geosciences (2003) *Geological Society, London, Special Publications*, 215 (1), pp. 1–6.

Menke H.P., Maes J., Geiger S. Upscaling the porosity–permeability relationship of a microporous carbonate for Darcy-scale flow with machine learning (2021) *Scientific Reports*, 11 (1), art. no. 2625.

Miarelli M., Della Torre A. Workflow development to scale up petrophysical properties from digital rock physics scale to laboratory scale (2021) *Transport in Porous Media*, 140 (2), pp. 459–492.

Nazari M.H., Tavakoli V., Rahimpour-Bonab H., Sharifi-Yazdi M. Investigation of factors influencing geological heterogeneity in tight gas carbonates, Permian reservoir of the Persian Gulf (2019) *Journal of Petroleum Science and Engineering*, 183, art. no. 106341.

Néri A., Guignard J., Monnereau M., Bystricky M., Perrillat J.-P., Andrault D., King A., Guignot N., Tenailleau C., Duployer B., Toplis M.J., Quitté G. Reevaluation of metal interconnectivity in a partially molten silicate matrix using 3D microtomography (2020) *Physics of the Earth and Planetary Interiors*, 308, art. no. 106571.

Nunes M., Valle B., Borghi L., Favoreto J., Mendes M. Multi-scale and multi-technique characterization of hybrid coquinas: A study case from the Morro do Chaves Formation (Barremian-Aptian of Sergipe-Alagoas Basin, Northeast Brazil) (2022) *Journal of Petroleum Science and Engineering*, 208, art. no. 109718.

Saraf S., Bera A. A review on pore-scale modeling and CT scan technique to characterize the trapped carbon dioxide in impermeable reservoir rocks during sequestration (2021) *Renewable and Sustainable Energy Reviews*, 144, art. no. 110986.

Saxena N., Hows A., Hofmann R., Freeman J., Appel M. Estimating pore volume of Rocks from pore-scale imaging (2019) *Transport in Porous Media*, 129 (1), pp. 403–412.

Tan D., Luo L., Song L., Liu F., Wang J., Gluyas J., Zhou H., Wang J., Zhu C., Mo S., Yu X., Tan X. Differential precipitation mechanism of cement and its impact on reservoir quality in tight sandstone: A case study from the Jurassic Shaximiao formation in the central Sichuan Basin, SW China (2023) *Journal of Petroleum Science and Engineering*, 221, art. no. 111263.

Tavakoli V. Corre Analysis: An Introduction (2018a) SpringerBriefs in Petroleum Geoscience and Engineering, pp. 1–13. In: Tavakoli V. *Geological core analysis: Application to reservoir characterization.* Springer, Cham, Switzerland, p. 99.

Tavakoli V. *Geological core analysis: Application to reservoir characterization* (2018b) Springer, Cham, Switzerland.

Tavakoli V., Rahimpour-Bonab H., Esrafili-Dizaji B. Diagenetic controlled reservoir quality of South Pars gas field, an integrated approach (2011) *Comptes Rendus Geoscience*, 343 (1), pp. 55–71.

Vinegar H.J. X-ray CT and NMR imaging of rocks (1986) *Journal of Petroleum Technology*, 38, pp. 257–259.

Vinegar H.J., Wellington S.L. Tomographic imaging of three-phase flow experiments (1986) *Review of Scientific Instruments*, 58, pp. 96–107.

Wang M., Wang R., Yuan S., Zhou F. A pore-scale study on the dynamics of spontaneous imbibition for heterogeneous sandstone gas reservoirs (2023) *Frontiers in Energy Research*, 11, art. no. 1135903.

Wang S., Tan M., Wang X., Zhang L. Microscopic response mechanism of electrical properties and saturation model establishment in fractured carbonate rocks (2022) *Journal of Petroleum Science and Engineering*, 208, art. no. 109429.

Yan Y.T., Chua S., DeCarlo T.M., Kempf P., Morgan K.M., Switzer A.D. Core-CT: A MATLAB application for the quantitative analysis of sediment and coral cores from X-ray computed tomography (CT) (2021) *Computers and Geosciences*, 156, art. no. 104871.

Yasuda E.Y., Koroishi E.T., Vidal Vargas J.A., Trevisan O.V. Dissolution evaluation of coquina, part 1: Carbonated-brine continuous injection using computed tomography and PHREEQC (2018) *Energy and Fuels*, 32 (4), pp. 5289–5301.

Yu C., Qin F., Li Y., Qin Z., Norell M. CT segmentation of dinosaur fossils by deep learning (2022) *Frontiers in Earth Science*, 9, art. no. 805271.

Zhao L., Liu F., Wang P., Liu P., Luo Z., Li N. A review of creation and propagation of complex hydraulic fracture network (2014) *Oil and Gas Geology*, 35 (4), pp. 562–569.

Zhao Z., Zhou X.P., Qian Q.H. DQNN: Pore-scale variables-based digital permeability assessment of carbonates using quantum mechanism-based machine-learning (2022) *Science China Technological Sciences*, 65 (2), pp. 458–469.

2

Facies and Diagenesis

2.1 Introduction

The final goal of reservoir evaluation is determining the amount of oil in place. Many properties are measured and interpreted for this purpose. All of them are controlled by geological characteristics. These include primary depositional facies or secondary diagenetic changes. Spatial distribution of reservoir properties is controlled by facies and sedimentary environments at a large scale while diagenesis controls the reservoir behavior at smaller scale. Many tools and measurements are used to evaluate the reservoir characteristics of the hydrocarbon bearing formations. Most of these methods are destructive and valuable samples are destroyed. CT scanning is one of the non-destructive methods which can be used with reasonable certainty. Various distinguishing properties of facies and different diagenetic impacts in these photos have been the center of considerable debate. This is more important in carbonate reservoirs with higher intrinsic heterogeneity. Therefore, this chapter considers the effect of these features in CT slices of carbonate cores. Routinely, the first data about the facies and diagenesis is gathered after core opening, plugging and trimming. Thin sections are prepared and microscopic geological attributes are documented. Then, cores are slabbed and macroscopic core description is performed. Identifying these features in carbonate cores prior to core opening and thin section preparation saves time and aids in the interpretation of facies, sedimentary environments, diagenetic processes and reservoir petrophysics. These sedimentary and diagenetic features also affect the digital petrophysics and its calculations.

Various types of lithology are observed in carbonate reservoirs. Calcite, dolomite and anhydrite are more common but fine clays are also present. This chapter considers the trace of various lithologies in CT scan images. Calcite, dolomite and anhydrite are different in density. This is also the case about the clay minerals. These fine siliciclastics have lowest density among these constituents. It continues with calcite, dolomite and anhydrite. Low-density minerals are darker in color while high-density lithologies are lighter. Therefore, lithologies are compared with each other (Figures 2.1–2.13).

DOI: 10.1201/9781003405412-2

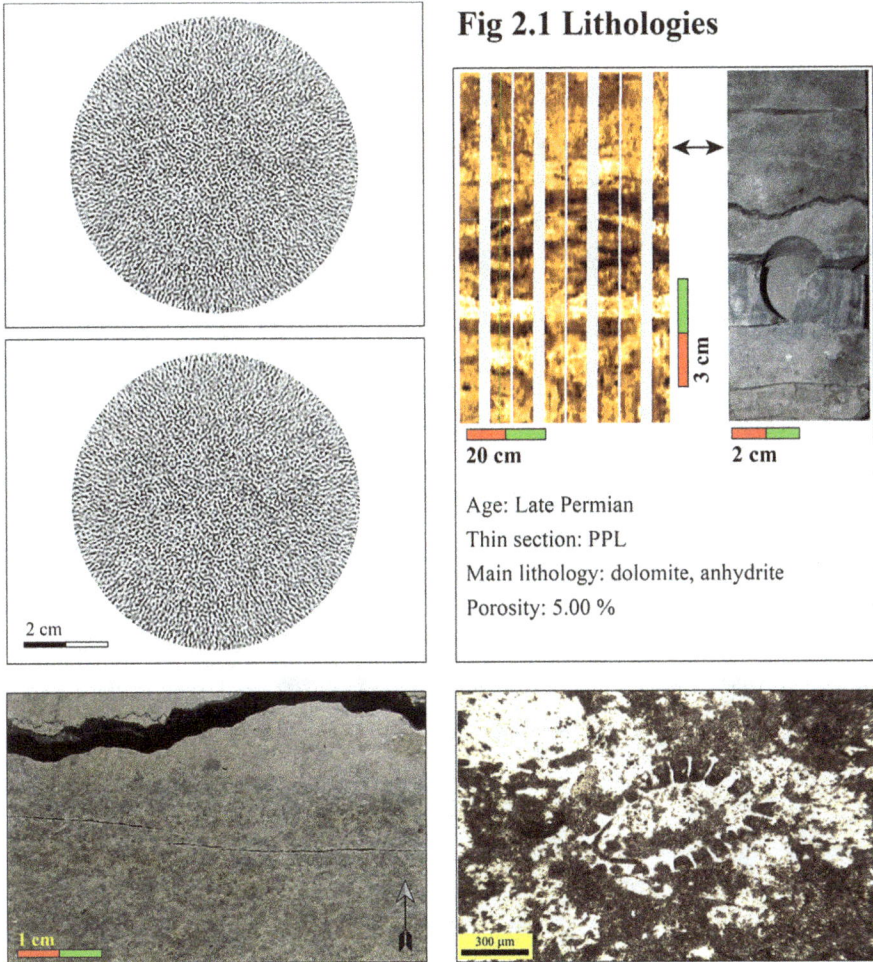

Fig 2.1 Lithologies

Age: Late Permian

Thin section: PPL

Main lithology: dolomite, anhydrite

Porosity: 5.00 %

FIGURE 2.1

In CT images, each mineral is identified by its gray color intensity. Change from dark to light gray color, indicates a change in mineralogy to denser mineral phases. Hence, mixture of minerals with various densities resulted in salt–pepper pattern in these images. Mixture of anhydrite and dolomite is evident in CT, close-up and thin section photos. These minerals are dispersed in the whole sample.

Fig 2.2 Lithologies

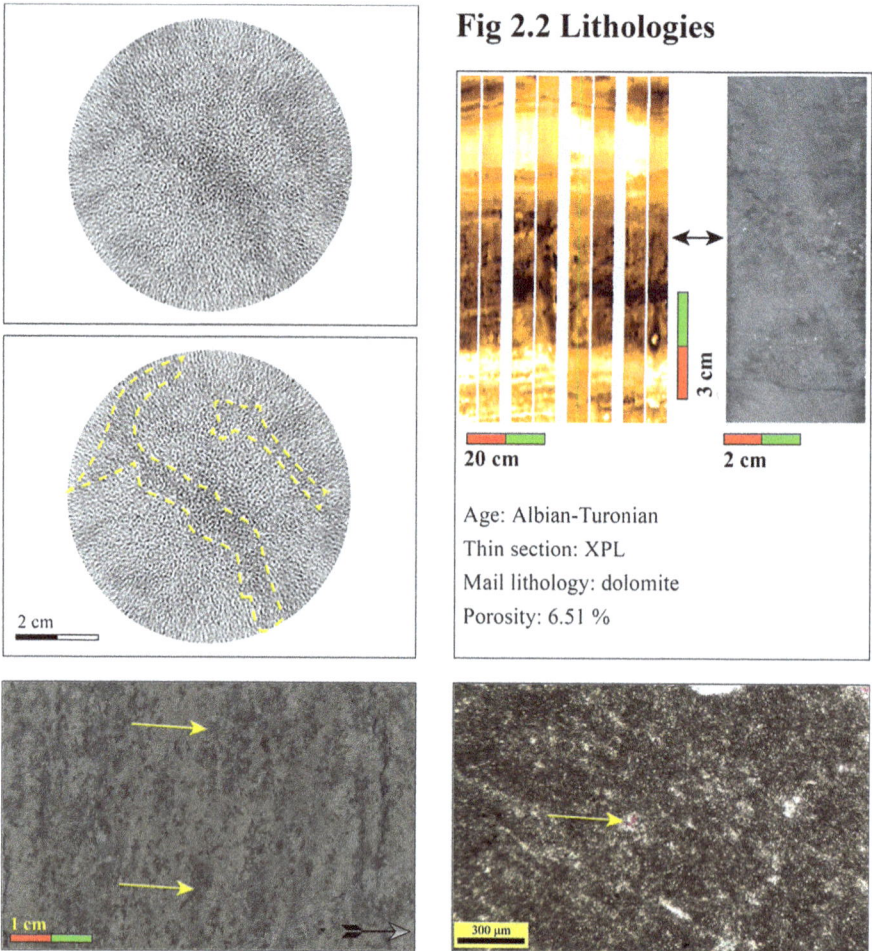

Age: Albian-Turonian
Thin section: XPL
Mail lithology: dolomite
Porosity: 6.51 %

FIGURE 2.2

Three types of lithology, including lime, dolomite and anhydrite are present in the images. This is evident from the changes in grayscale color in the CT image (dashed parts), change in the color of the whole core photo and combination of dark and light colors in FMI image. Layering in lithology is obvious in all photos. Thin layers can also be distinguished in thin section studies. Compaction can be inferred from thin dark layers of the whole core photo and FMI log. Yellow arrows show minor amounts of anhydrite.

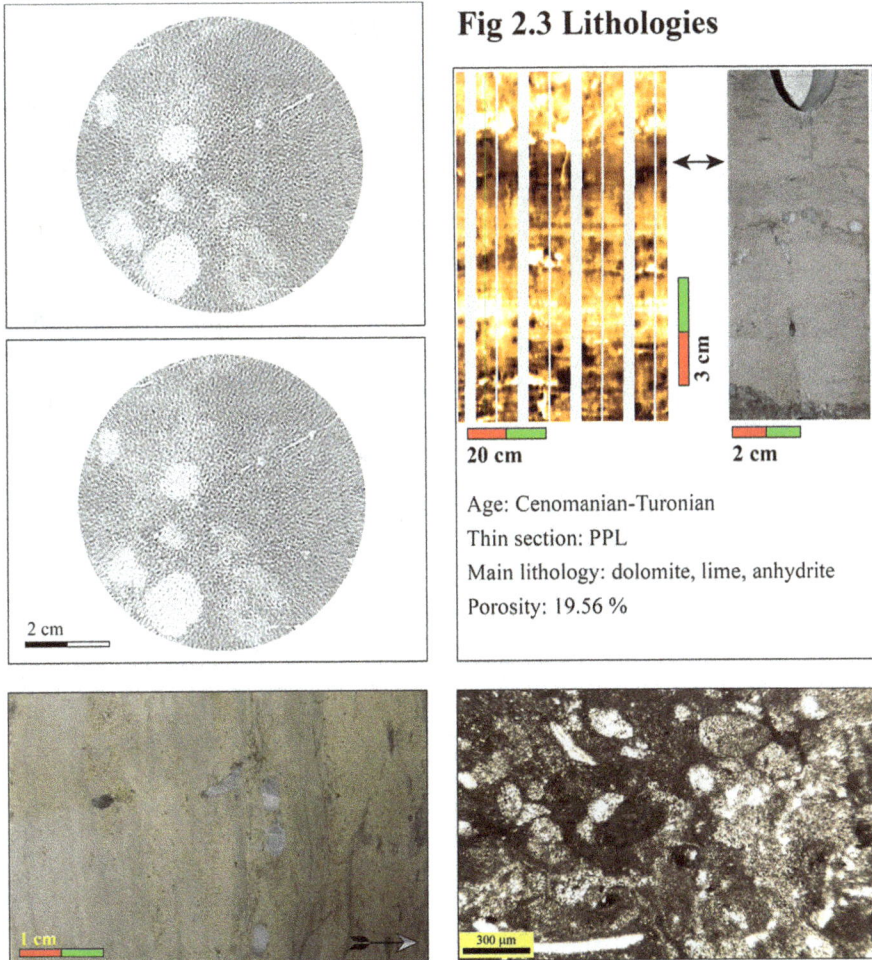

Fig 2.3 Lithologies

20 cm   2 cm

Age: Cenomanian-Turonian
Thin section: PPL
Main lithology: dolomite, lime, anhydrite
Porosity: 19.56 %

FIGURE 2.3

Dolomite (60 %), lime (30 %) and anhydrite (10 %) are present in this sample and therefore it is suitable for comparison of different lithologies in CT images. Anhydrite nodules show lighter color than others because of their higher density. Lime is recognized by the darkest color. Dolomites are gray. A vertical filled fracture is also visible in CT, FMI and whole core photos.

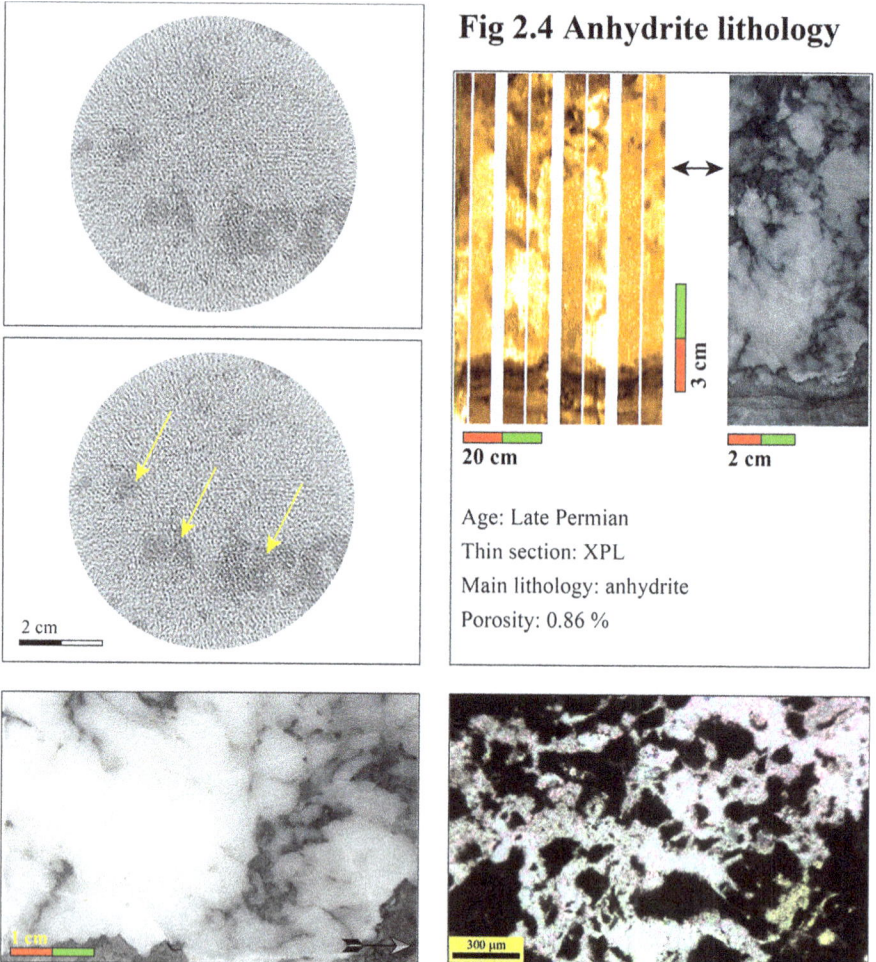

Fig 2.4 Anhydrite lithology

Age: Late Permian
Thin section: XPL
Main lithology: anhydrite
Porosity: 0.86 %

FIGURE 2.4

Anhydrite mineral is present in various forms and shapes (e.g., nodules, layers, cements) in carbonate reservoirs. They are characterized by light color in CT images because of their high density compared to carbonates (calcite and dolomite). Other minerals (specially dolomite) with dark shade of gray are also visible in these images (yellow arrows).

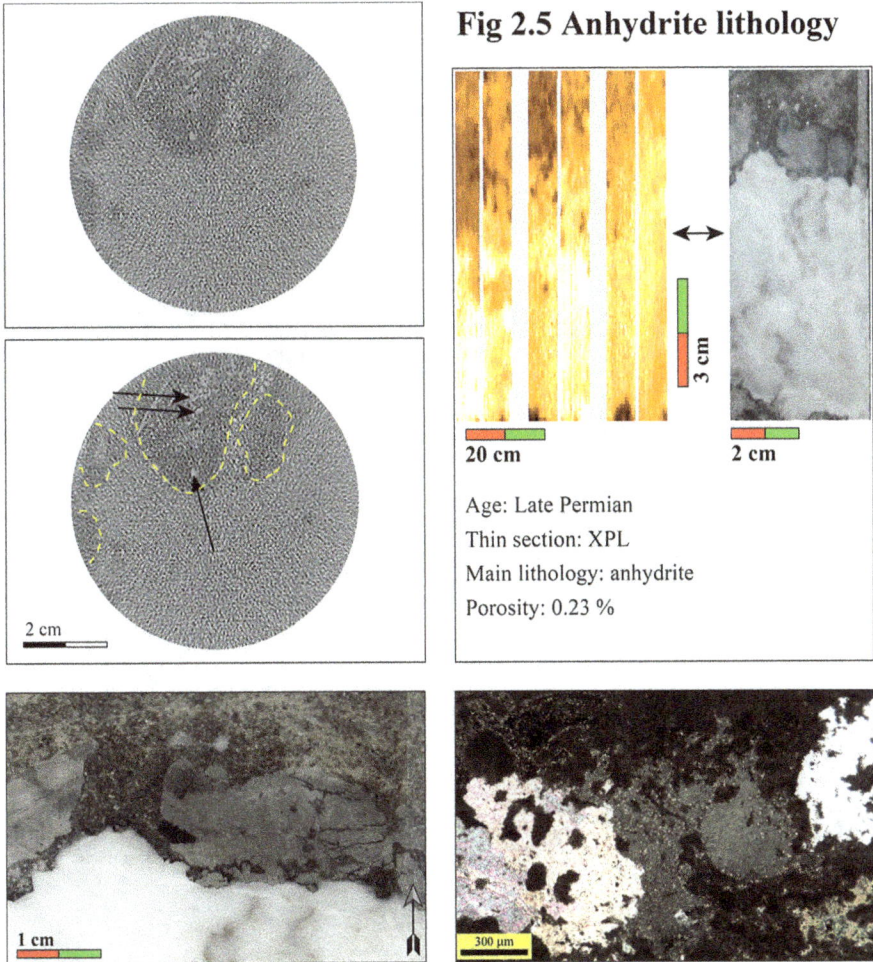

Fig 2.5 Anhydrite lithology

Age: Late Permian
Thin section: XPL
Main lithology: anhydrite
Porosity: 0.23 %

FIGURE 2.5

Anhydrite reduces the reservoir quality of carbonate formations. Therefore, it is important to recognize them using various proxies. Here, a mixture of dolomite (dashed yellow perimeters) and anhydrites can be seen in the CT slice. They are marked by white color in both CT and FMI images. In many cases, inclusions of anhydrites are present in the dolomitic parts (black arrows).

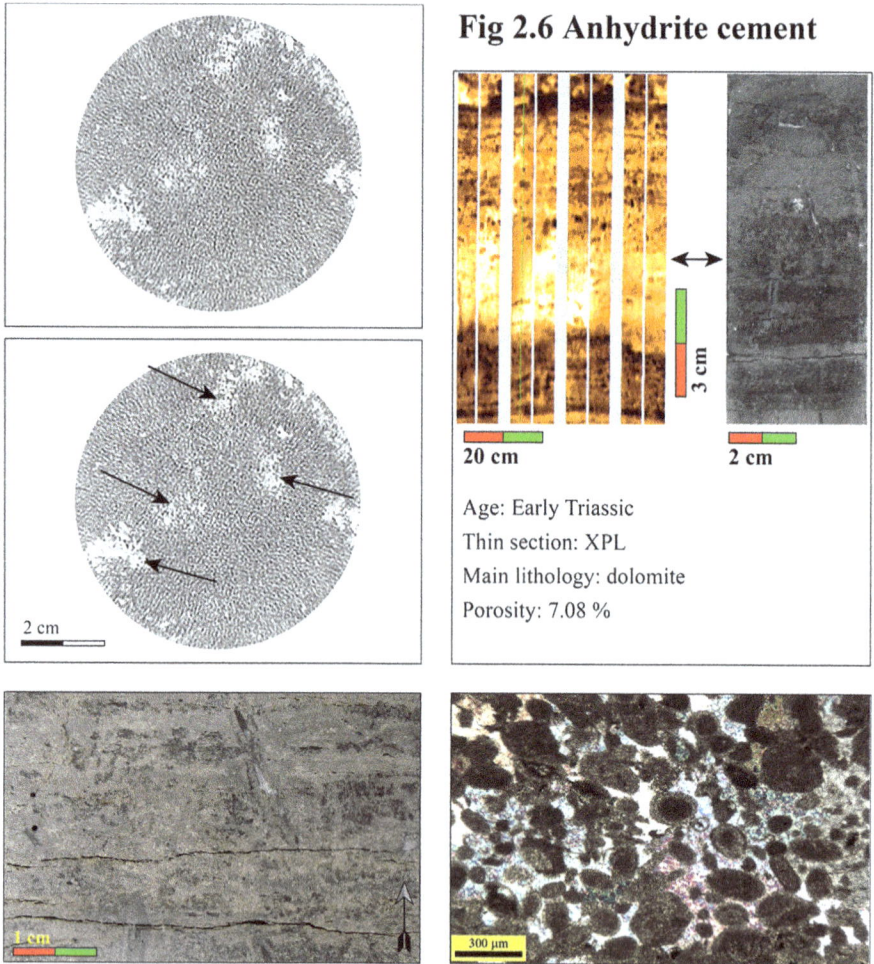

Fig 2.6 Anhydrite cement

Age: Early Triassic

Thin section: XPL

Main lithology: dolomite

Porosity: 7.08 %

FIGURE 2.6

Anhydrite is recognized by light color in carbonate CT scan images because of its higher density compared to calcite and dolomite. Its shape and type of distribution indicate its form in these images. The spotty distribution and jagged margins show anhydrite cements (black arrows) in this sample. These anhydrites are also visible in whole core, close-up, thin section and FMI image.

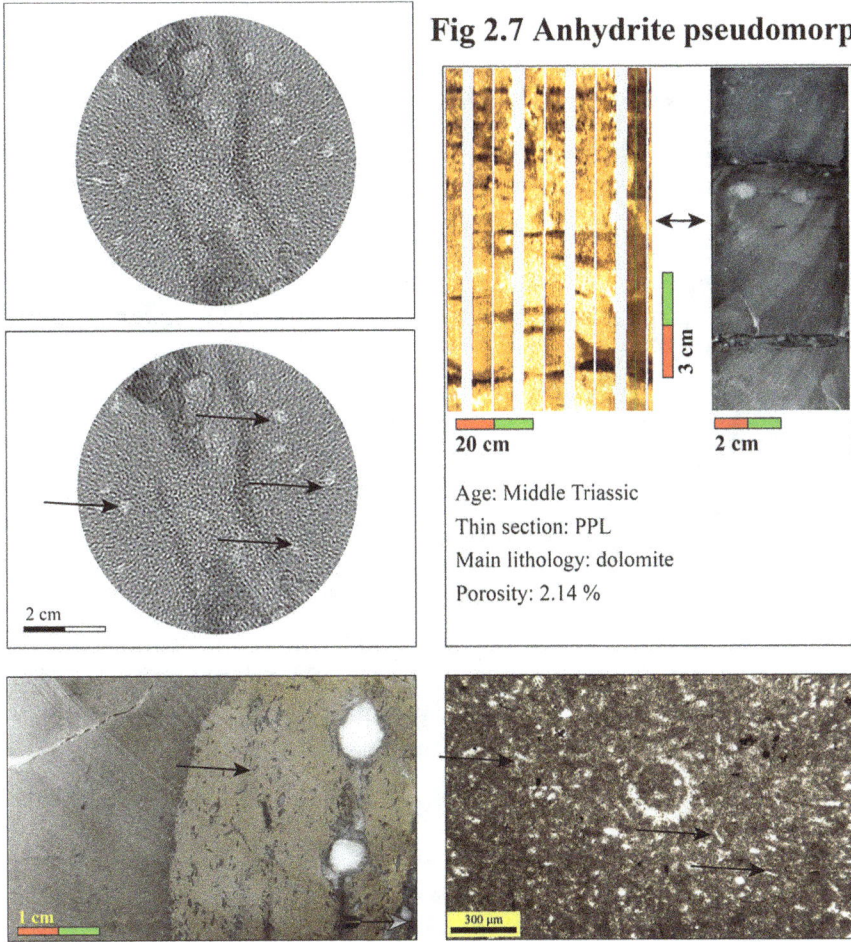

Fig 2.7 Anhydrite pseudomorph

Age: Middle Triassic
Thin section: PPL
Main lithology: dolomite
Porosity: 2.14 %

FIGURE 2.7
Gypsum is transformed to anhydrite in burial diagenesis, but the mineral retains its original shape. The resulted feature is called pseudomorph. Regarding higher anhydrite density compared to carbonate matrix (calcite or dolomite), it can be distinguished by its lighter color in CT scan image (black arrows). These pseudomorphs are also visible in core and thin section studies.

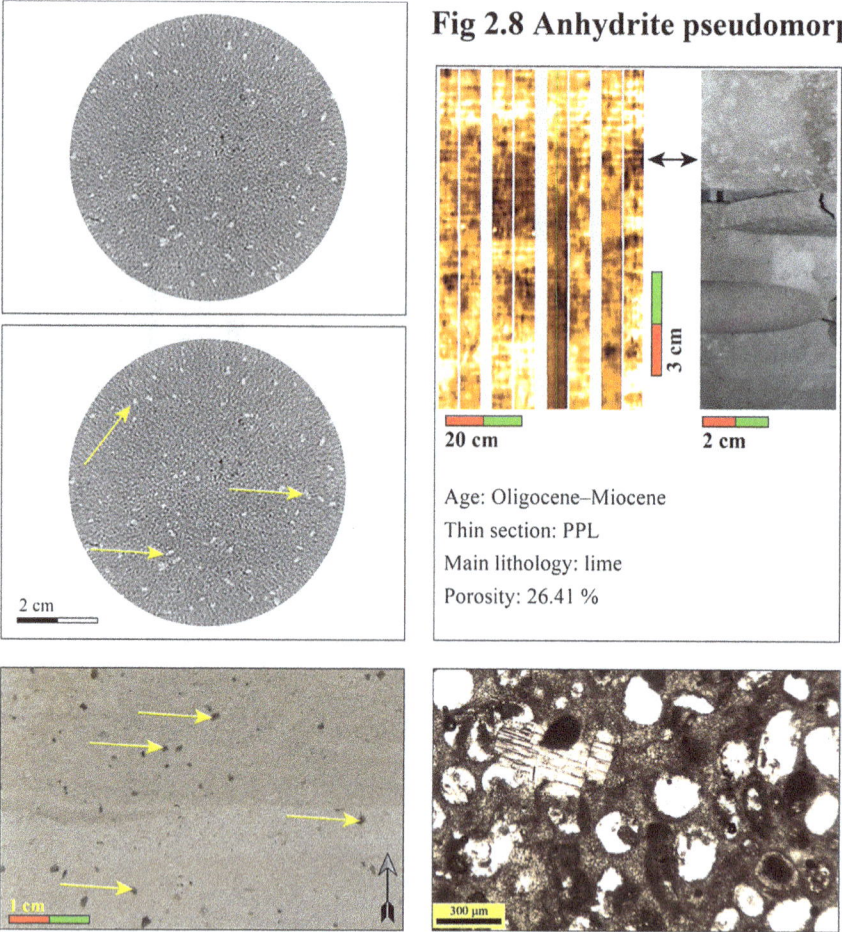

Fig 2.8 Anhydrite pseudomorph

20 cm

2 cm

Age: Oligocene–Miocene

Thin section: PPL

Main lithology: lime

Porosity: 26.41 %

2 cm

1 cm

300 µm

FIGURE 2.8

Anhydrite pseudomorphs after gypsum crystals preserve the original shape of the mineral and indicate the various stages of diagenetic processes of carbonate formations. They are easily recognized by the crystal shape, light color and dispersion in the matrix in CT images (yellow arrows). The sample is limestone and so there is enough contrast in the image. Some moldic pores are also visible with dark gray color.

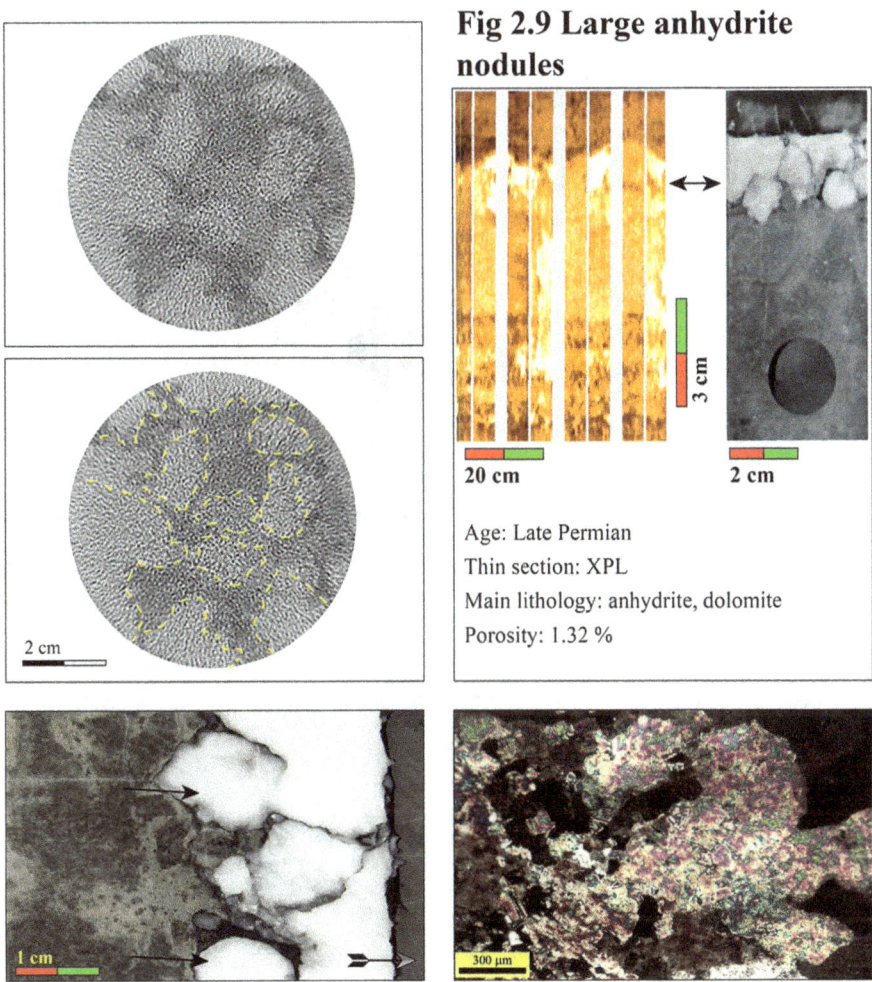

Fig 2.9 Large anhydrite nodules

Age: Late Permian
Thin section: XPL
Main lithology: anhydrite, dolomite
Porosity: 1.32 %

FIGURE 2.9
Large anhydrite nodules with high density are evident with their light color. The matrix is dolomite and so the contrast is not so clear. These anhydrites are routinely formed in primary depositional environments in carbonate-evaporite series. They are recognizable on core, FMI and thin section studies.

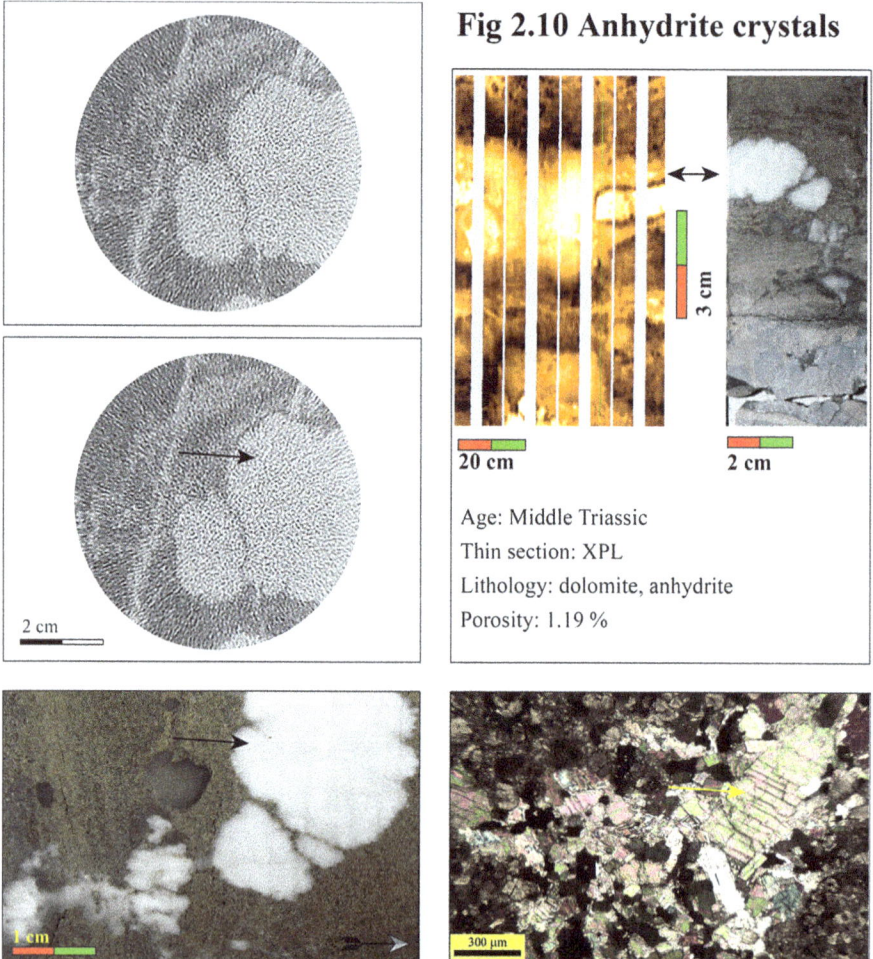

Fig 2.10 Anhydrite crystals

Age: Middle Triassic

Thin section: XPL

Lithology: dolomite, anhydrite

Porosity: 1.19 %

FIGURE 2.10

Anhydrites are formed with different textures in sedimentary rocks. The matrix is dolomite or limestone in most cases and so they are lighter than the background in the CT images. Anyway, they show a salt–pepper pattern when large anhydrite crystals have been formed in the sedimentary environment or later diagenetic processes (black arrow). In this case, main lithologies are dolomite and anhydrite, which create different parts with high contrast in CT image.

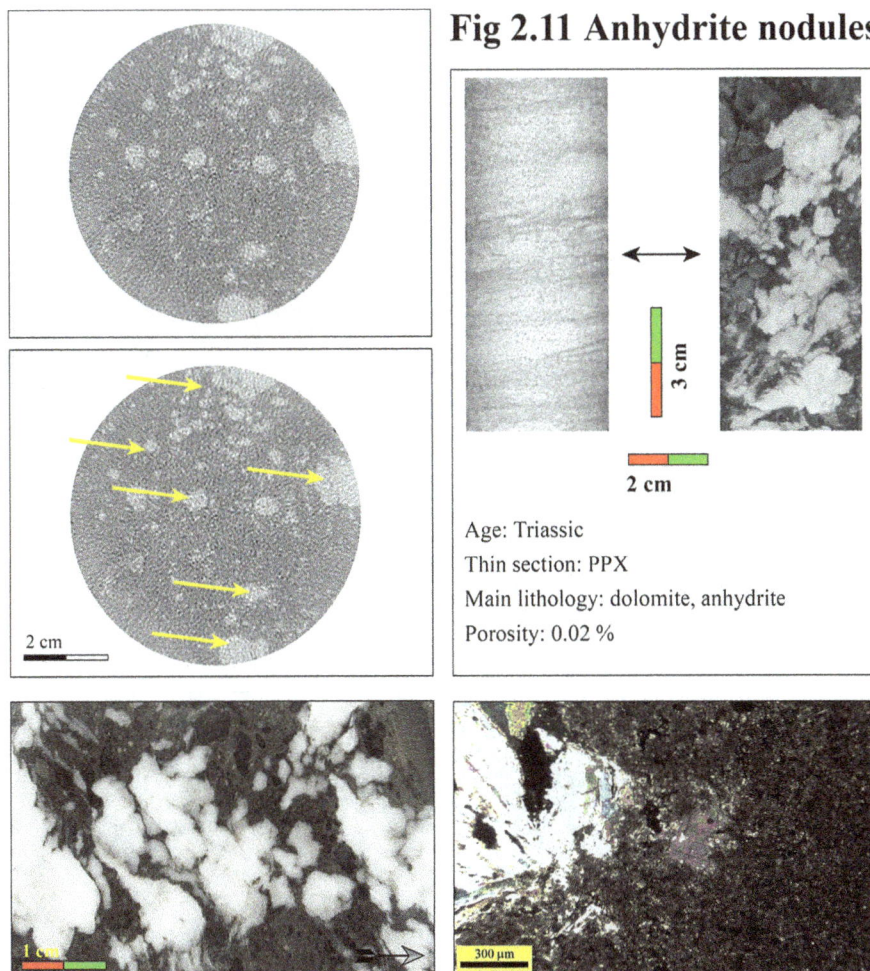

Fig 2.11 Anhydrite nodules

Age: Triassic
Thin section: PPX
Main lithology: dolomite, anhydrite
Porosity: 0.02 %

FIGURE 2.11
Frequent anhydrite nodules are obvious in the dolomitic matrix. Anhydrite with higher density is easily distinguishable on CT images with lighter color (milky white). Cross laminations are visible on longitudinal CT image but cannot be seen on transvers CT photos. Anhydrites have occluded the pore spaces and porosity has been reduced. This is evident from low porosity of the sample.

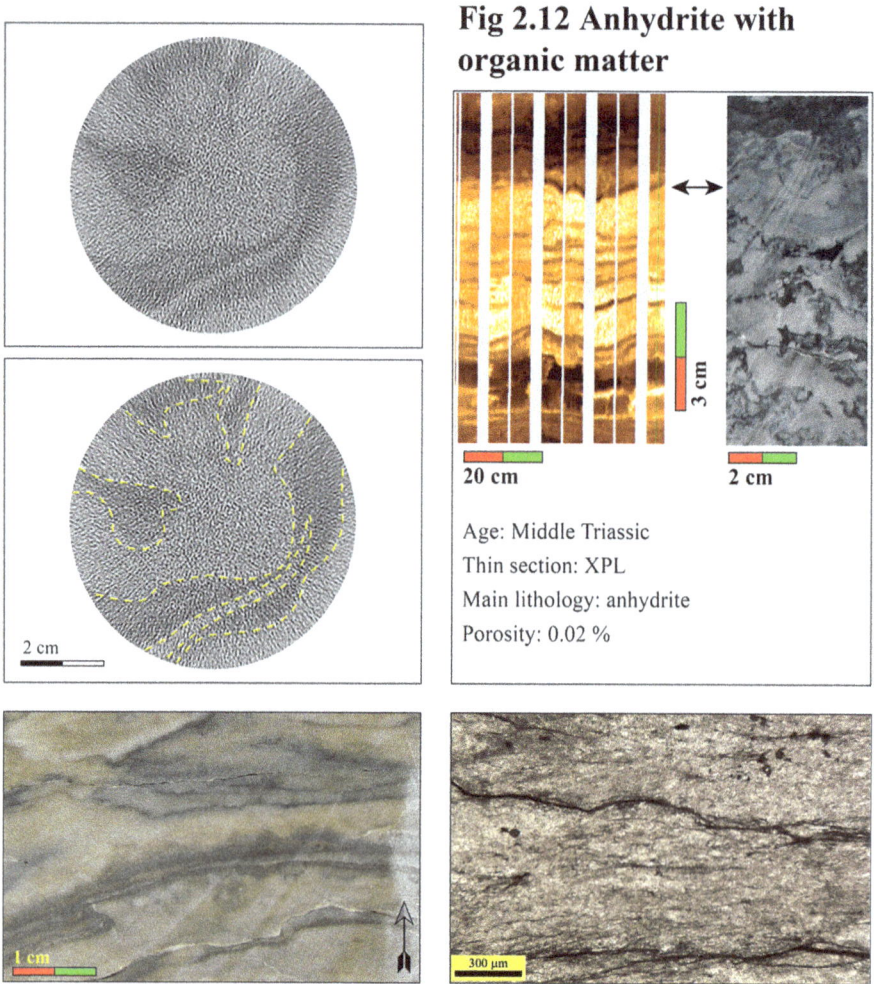

Fig 2.12 Anhydrite with organic matter

Age: Middle Triassic
Thin section: XPL
Main lithology: anhydrite
Porosity: 0.02 %

FIGURE 2.12
A plenty of microorganisms live in the shallow depositional environments. These environments are also susceptible to precipitation of evaporites. Therefore, alternations of evaporites (specially anhydrite) and organic materials are frequently seen in peritidal (mostly sabkha) depositional settings. Large difference in density clearly marks these layers in CT images. Anhydrites with high density are white, while low-density organic materials are dark gray.

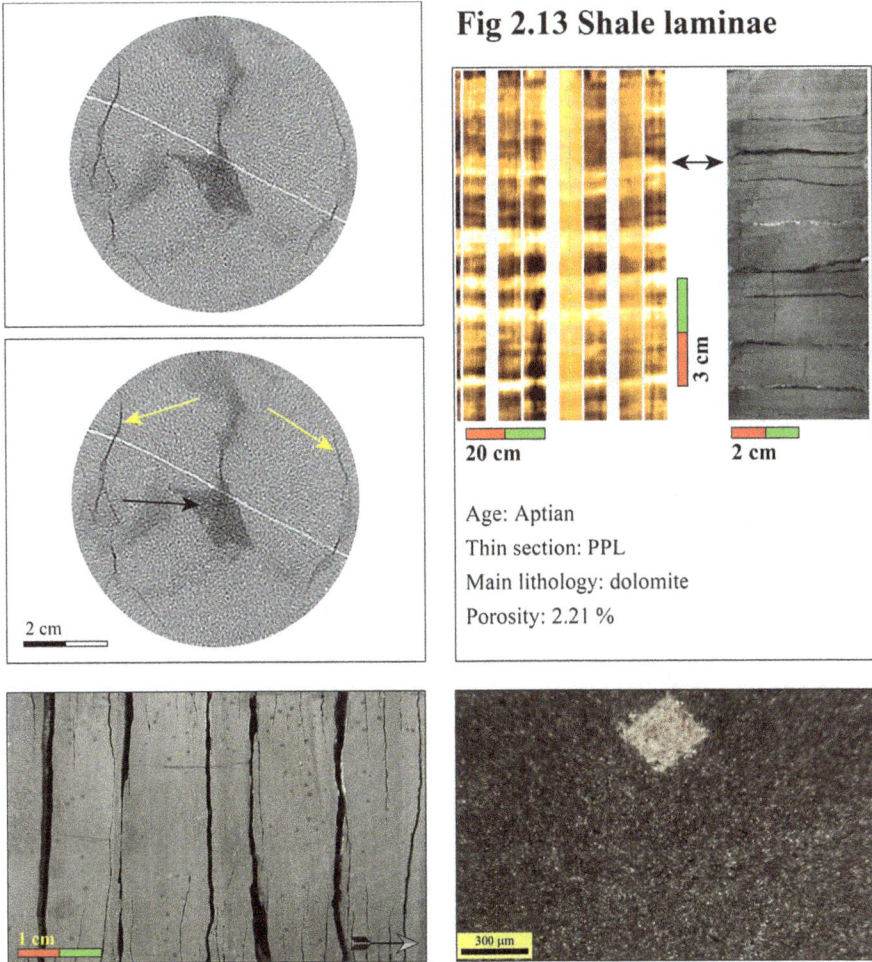

Fig 2.13 Shale laminae

Age: Aptian

Thin section: PPL

Main lithology: dolomite

Porosity: 2.21 %

FIGURE 2.13

Clays are phyllosilicate minerals that have terrestrial origin in many cases. However, they are sometimes present in carbonate environments. Lamination of these rocks is due to their mineral's structure and depositional mechanism. These laminations are visible in CT images as dark lines, almost parallel to each other (yellow arrows). Laminations are also visible on FMI image. High organic content is one of their routine characteristics (black arrow).

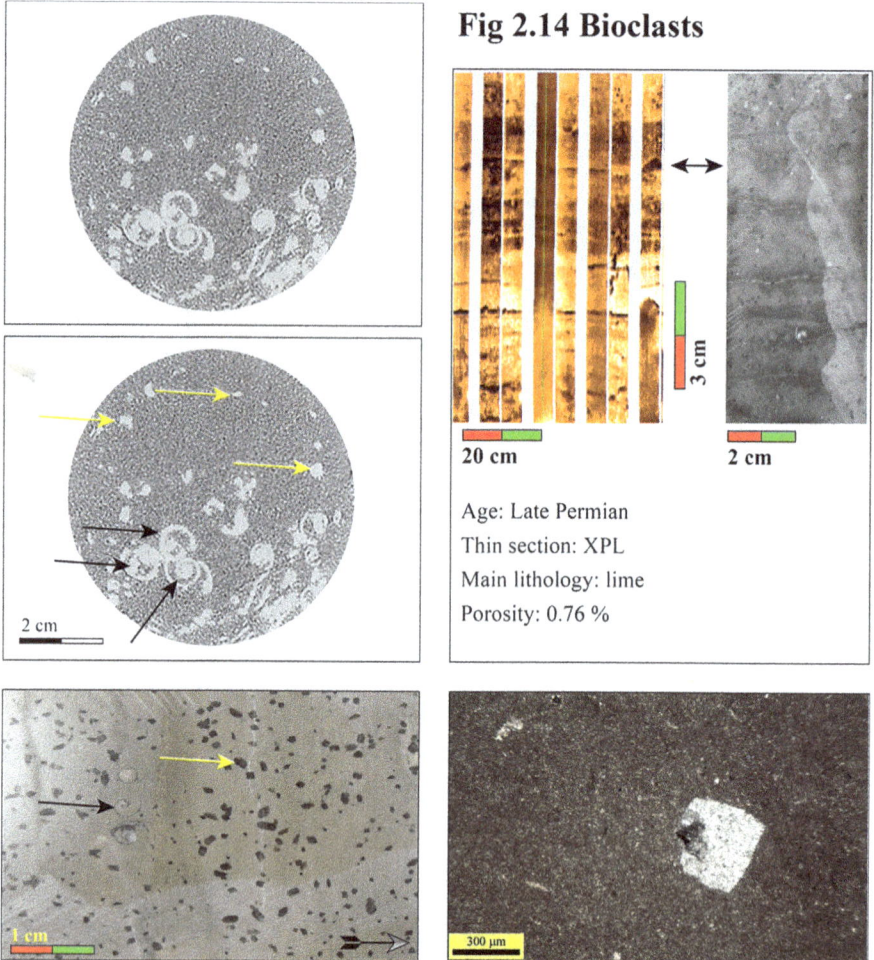

Fig 2.14 Bioclasts

20 cm   2 cm

Age: Late Permian
Thin section: XPL
Main lithology: lime
Porosity: 0.76 %

FIGURE 2.14

Large fossils in muddy matrix are seen in this CT image. Increasing evaporation caused death of the organisms. This is evident from frequent anhydrite pseudomorphs in the sample. Both bioclasts (black arrow) and anhydrite crystals (yellow arrow) are visible in CT image. They are also evident in core studies. These features are not large enough to be detected on FMI images.

Different anhydrite textures are also illustrated in this chapter. There is also one figure showing shale lithology. This type of lithology is not common in carbonates but it can be distinguished by dark color and fine laminations in CT slices. The chapter continues with consideration of main allochems, specially bioclasts and intraclasts, and how they are seen in these pictures. Different types and sizes of these allochems are presented in this chapter (Figure 2.14–2.23). These allochems along with the amount of micrite

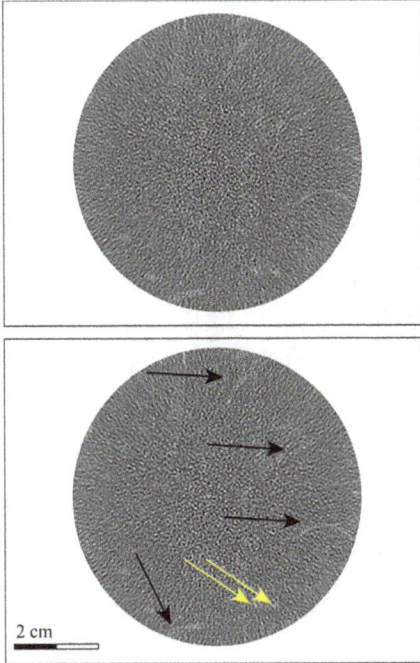

Fig 2.15 Large bioclasts

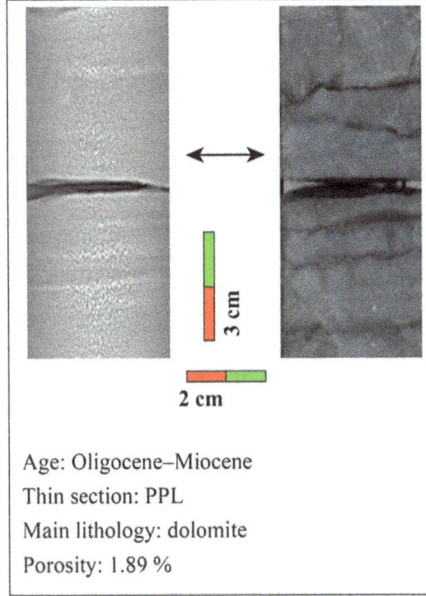

Age: Oligocene–Miocene
Thin section: PPL
Main lithology: dolomite
Porosity: 1.89 %

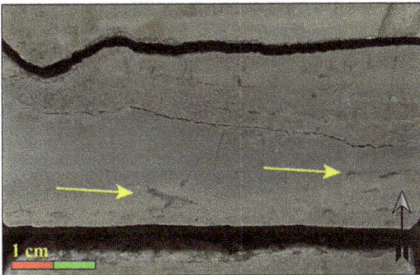

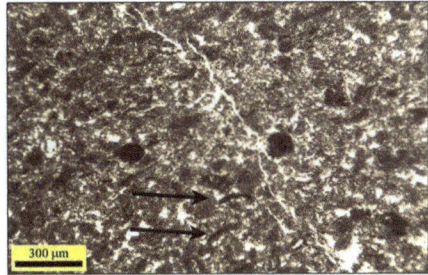

FIGURE 2.15

Large bioclasts are obvious in CT slice, whole-core photo, close-up and thin section images. Based on mineralogy (calcite, aragonite, dolomite and anhydrite), they are lighter or darker than the matrix in CT image. Some bioclasts are distinguishable by their original shape (black arrows) and some others have been broken in the depositional environment (yellow arrows).

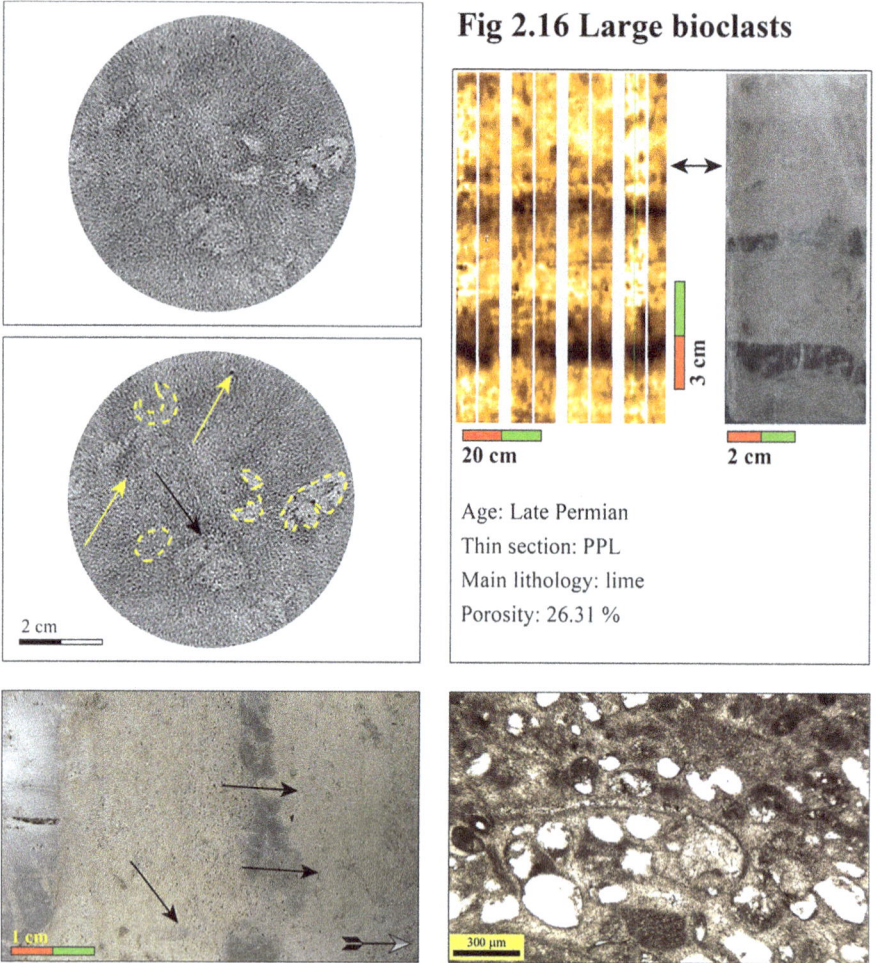

Fig 2.16 Large bioclasts

20 cm   2 cm

Age: Late Permian
Thin section: PPL
Main lithology: lime
Porosity: 26.31 %

FIGURE 2.16

Large bioclasts are obvious by lighter colors in the CT slice. Micrites appear darker than calcite crystals because of their lower density (high microporosity). Therefore, crystalized shells are lighter than surrounding micrite or microspar matrix. Micrite envelope is evident by darker color (black arrow on CT image). Porosities are also visible (yellow arrows).

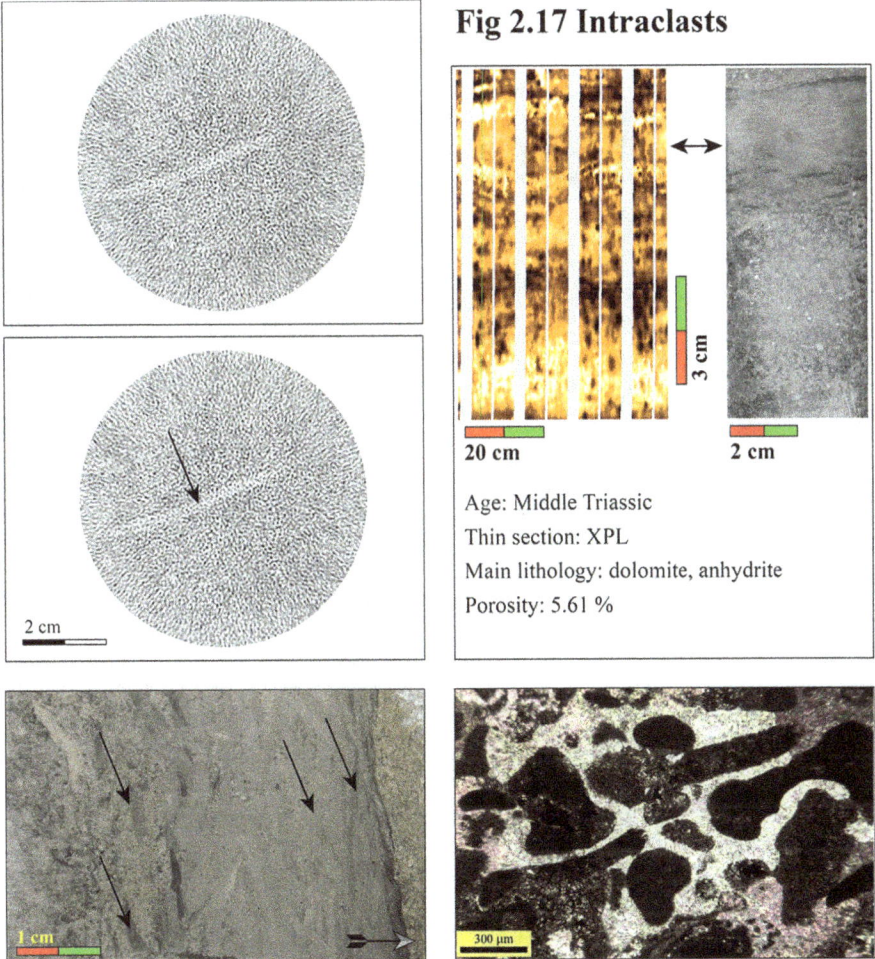

Fig 2.17 Intraclasts

Age: Middle Triassic

Thin section: XPL

Main lithology: dolomite, anhydrite

Porosity: 5.61 %

FIGURE 2.17

Horizontal intraclasts are seen as elongated objects in CT images. They show an increase in environmental energy and therefore are very important for interpretation of ancient sedimentary environments. They can be seen as light or dark objects, depending on difference between their and the matrix densities. Cross-bedding is visible in whole core photo. It is also traceable in FMI images (as a sinusoidal line).

Fig 2.18 Large intraclasts

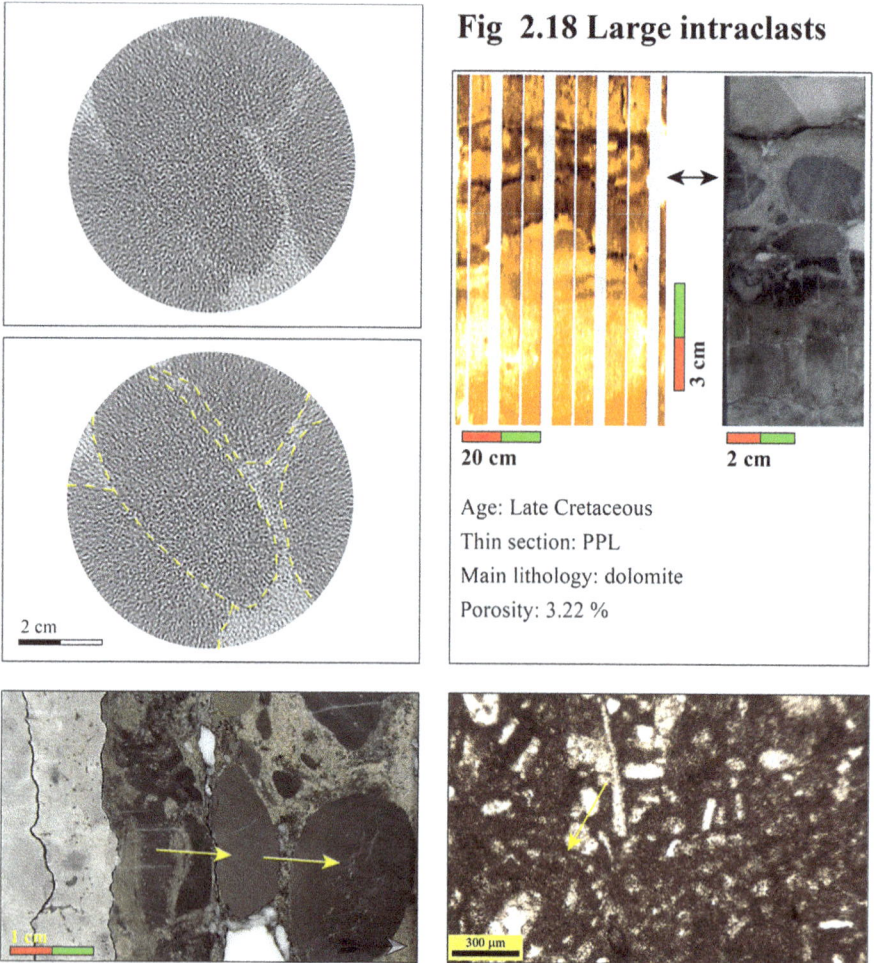

20 cm   2 cm

Age: Late Cretaceous
Thin section: PPL
Main lithology: dolomite
Porosity: 3.22 %

2 cm

1 cm

300 µm

FIGURE 2.18

Intraclasts show reactivation of depositional energy or just intrabasinal reworking. In this case, they are composed of limestone in dolomite background and so can be recognized by their darker color. The roundness of these intraclasts indicates high energy of environmental settings at the time of deposition. Large intraclasts serve as good indicators for distinguishing sequence stratigraphic surfaces.

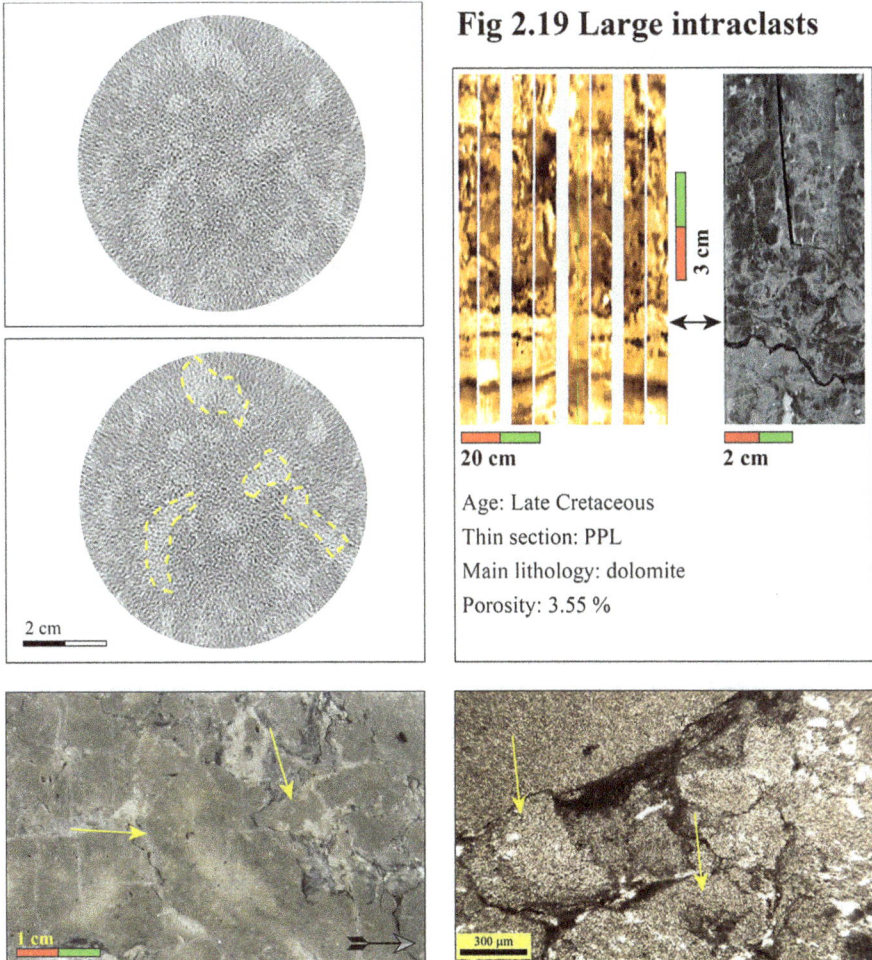

Fig 2.19 Large intraclasts

Age: Late Cretaceous
Thin section: PPL
Main lithology: dolomite
Porosity: 3.55 %

FIGURE 2.19

Intraclasts have been eroded from older sediments and redeposited near their provenance. With higher density and anhydrite content, these allochems are lighter than surrounding dolomites. They originated from proximal areas with high rate of evaporation. Therefore, the changes in grey intensity indicate a change in mineralogy to denser mineral phases. Large clasts are also visible in FMI image.

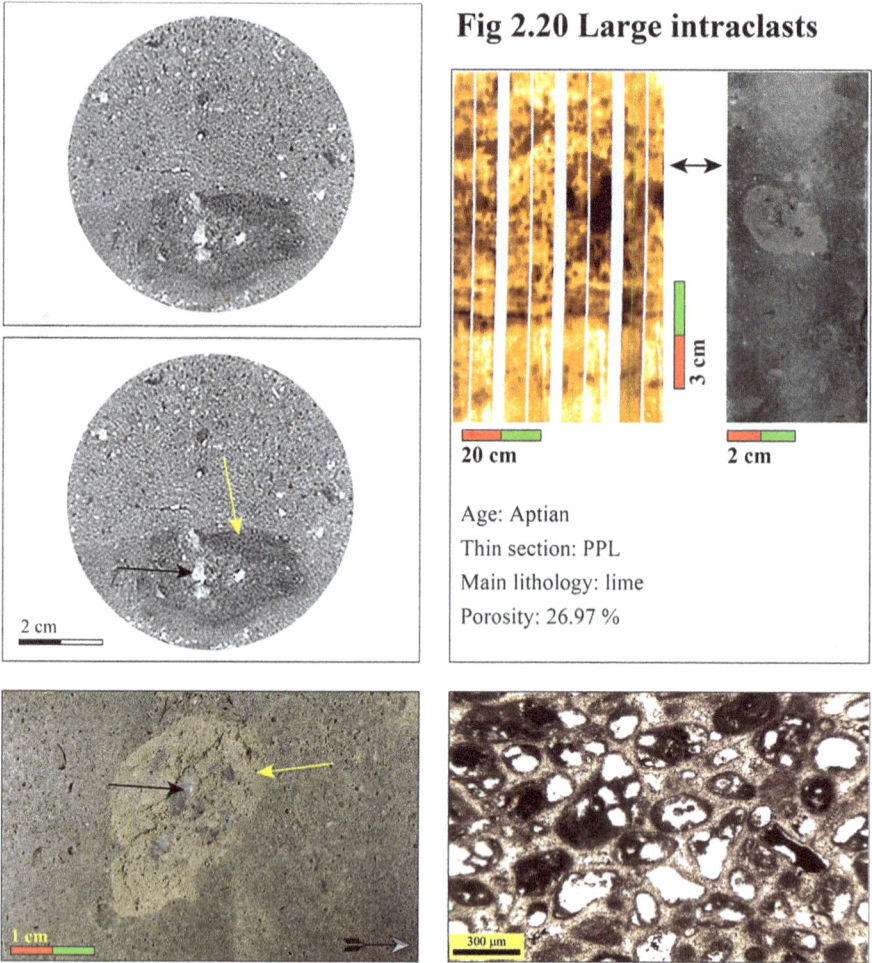

Fig 2.20 Large intraclasts

Age: Aptian
Thin section: PPL
Main lithology: lime
Porosity: 26.97 %

FIGURE 2.20

Intraclasts are made of micrites in many cases. Hence, they are marked by dark color in calcite or dolomite background. A large intraclast is visible in this sample (yellow arrows) in CT images, whole core photo, FMI log and close-up image. Intraclasts are routinely reworked from shallow environments and so contain anhydrite nodules (black arrows) in some cases.

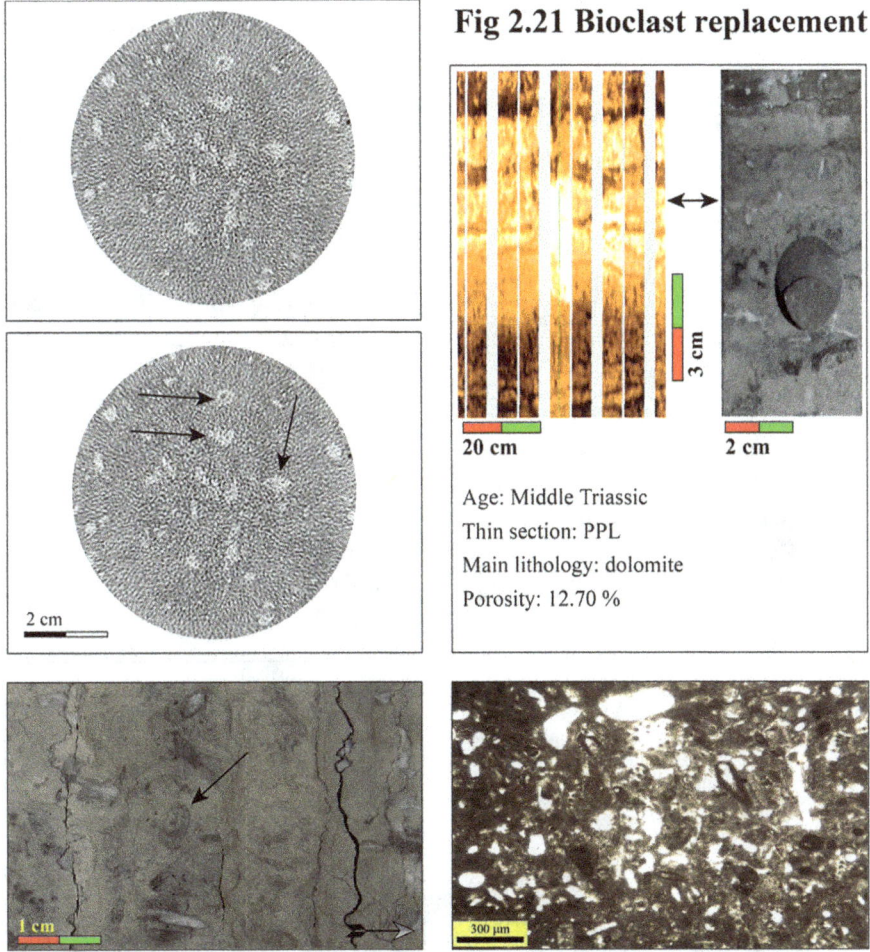

Fig 2.21 Bioclast replacement

Age: Middle Triassic
Thin section: PPL
Main lithology: dolomite
Porosity: 12.70 %

FIGURE 2.21
Bioclasts have been dissolved in meteoric depositional setting and have been filed by anhydrite in later diagenetic stages. The anhydrite mineral is characterized by its white color in CT and FMI images. Original shapes of some bioclasts are still recognizable (black arrows). Bioclasts, their filled casts and peloids are observable in thin section photo. Presence of different allochems caused high contrast between various parts of the CT image.

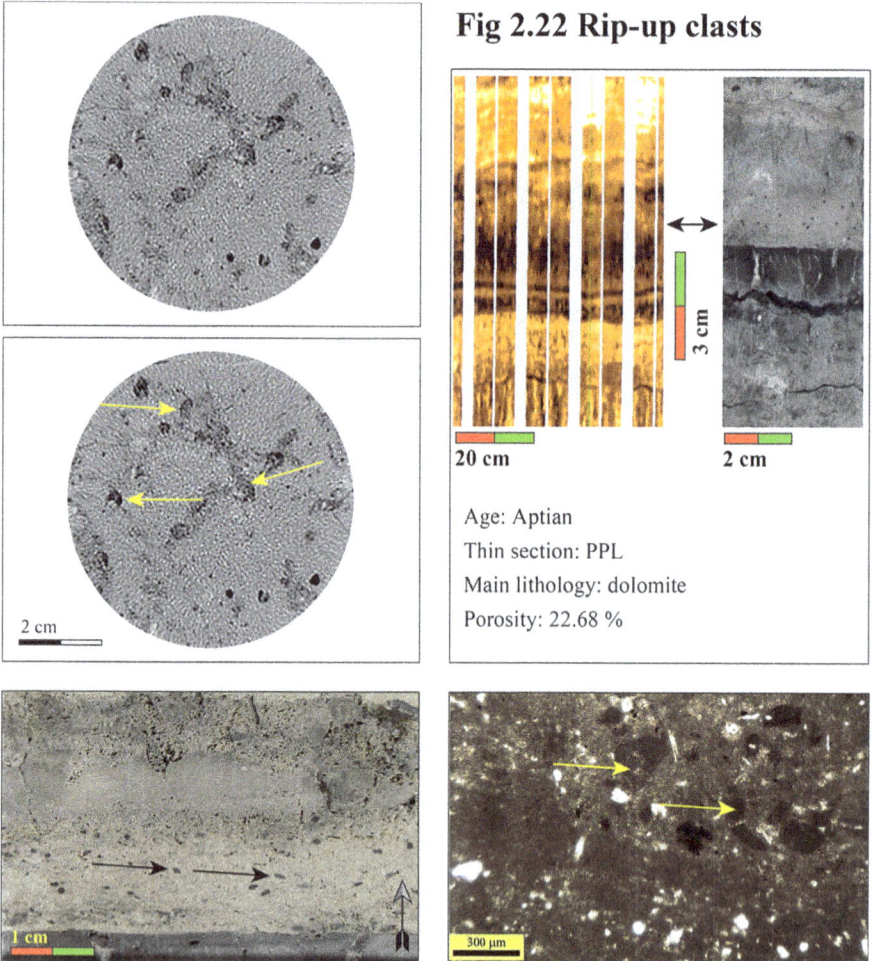

Fig 2.22 Rip-up clasts

20 cm   2 cm

Age: Aptian

Thin section: PPL

Main lithology: dolomite

Porosity: 22.68 %

2 cm

FIGURE 2.22

Rip-up clasts show re-activation of the depositional energy. They are irregular shaped clasts that have been detached from lower layer and have been deposited in the upper one. The lower layer is routinely formed by clay-dominated facies, because it has been deposited in the low energy level before re-activation. Therefore, the clasts are recognizable by higher gray intensity and darker color in CT images (arrows). Desiccation cracks are visible on whole core image.

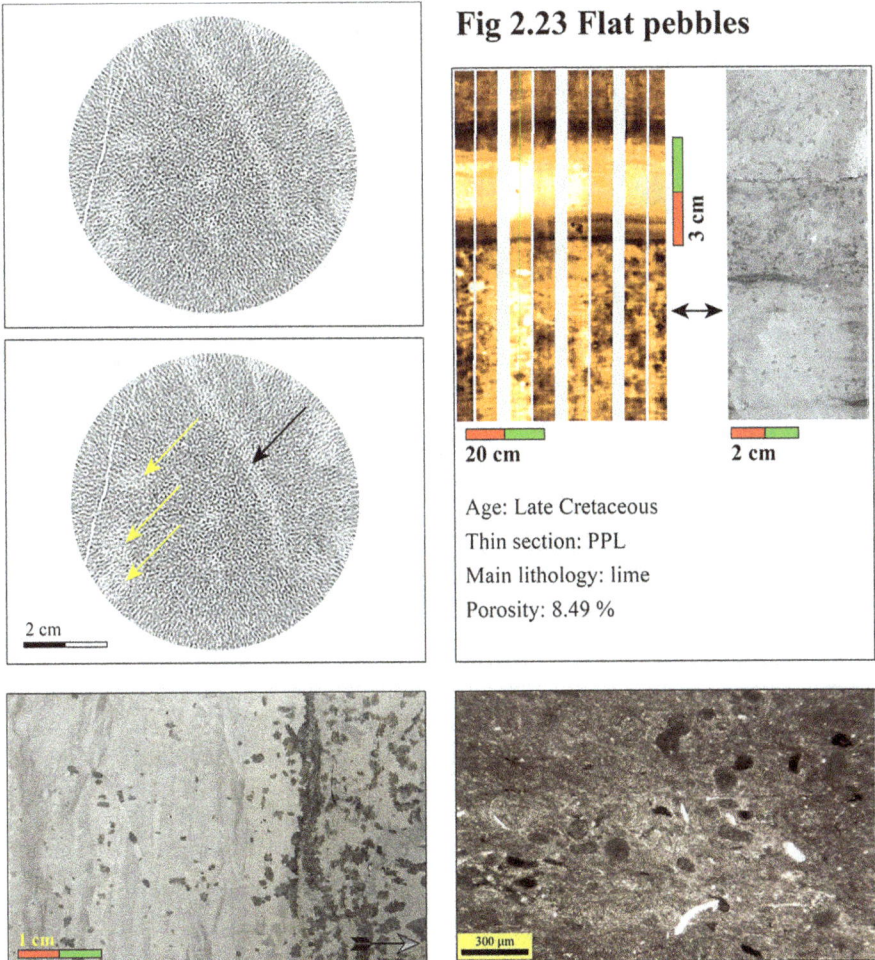

Fig 2.23 Flat pebbles

Age: Late Cretaceous
Thin section: PPL
Main lithology: lime
Porosity: 8.49 %

FIGURE 2.23

Flat pebbles are produced by reworking of previously deposited sediments. In this sample, they content more calcite crystals and so are marked by much lighter color, compared to the micritic background. Depending on the pebble and CT slice orientations, they can be traced as elongated (black arrow) or circle (yellow arrow) surfaces on CT images. These clasts indicate variations in depositional energy.

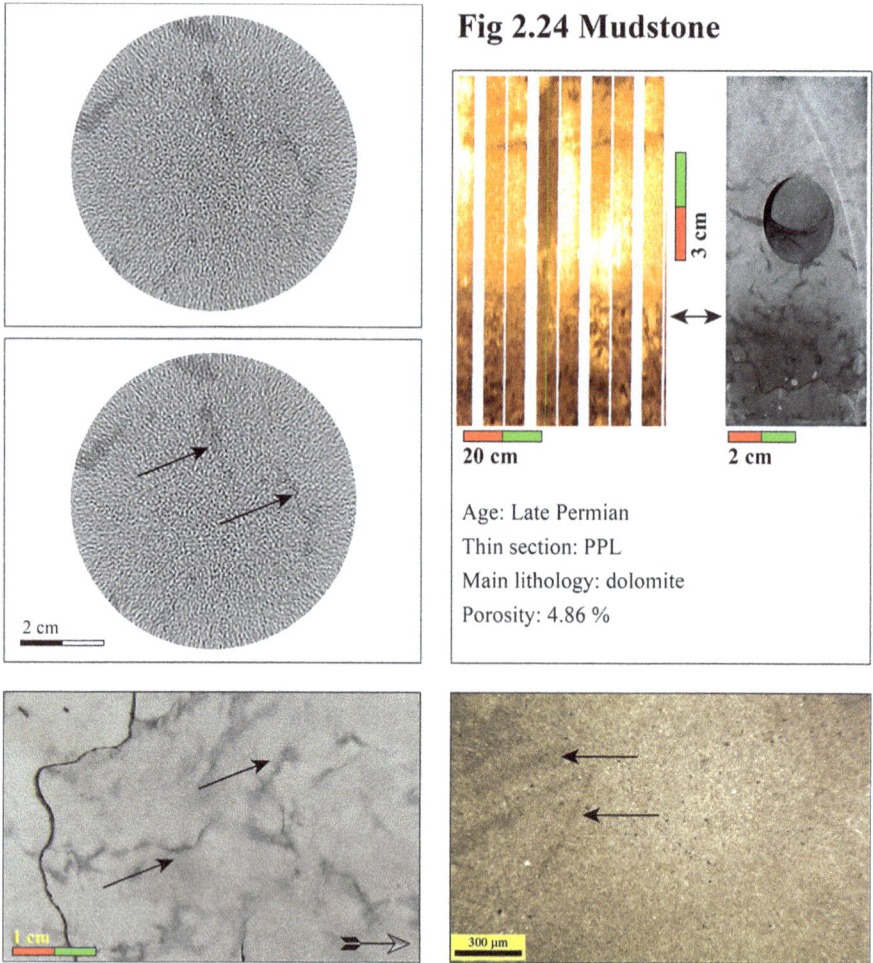

Fig 2.24 Mudstone

Age: Late Permian
Thin section: PPL
Main lithology: dolomite
Porosity: 4.86 %

FIGURE 2.24
A mudstone sample with high organic content can be seen in these images. They are marked by jagged dark lines on CT images. The traces of these materials are not continuous, even at core scale, and so can be distinguished from stylolites or solution seams. These organic matters are also recognizable in thin section, close-up and core photos (black arrows), as well as FMI log.

determine the texture of the carbonate rocks. The texture is very important for interpretation of physical, chemical and biological conditions of the sedimentary environments. Therefore, carbonate textures are considered after the allochems (Figures 2.24–2.40). Boundstone, mudstone, wackestone, packstone and grainstone textures with various characteristics are observed in these figures. There is also one figure showing crystalline carbonate texture with dolomite lithology. All of these textures may have the same mineralogy and so they are recognized through other factors, such as type of allochems and degree of heterogeneity.

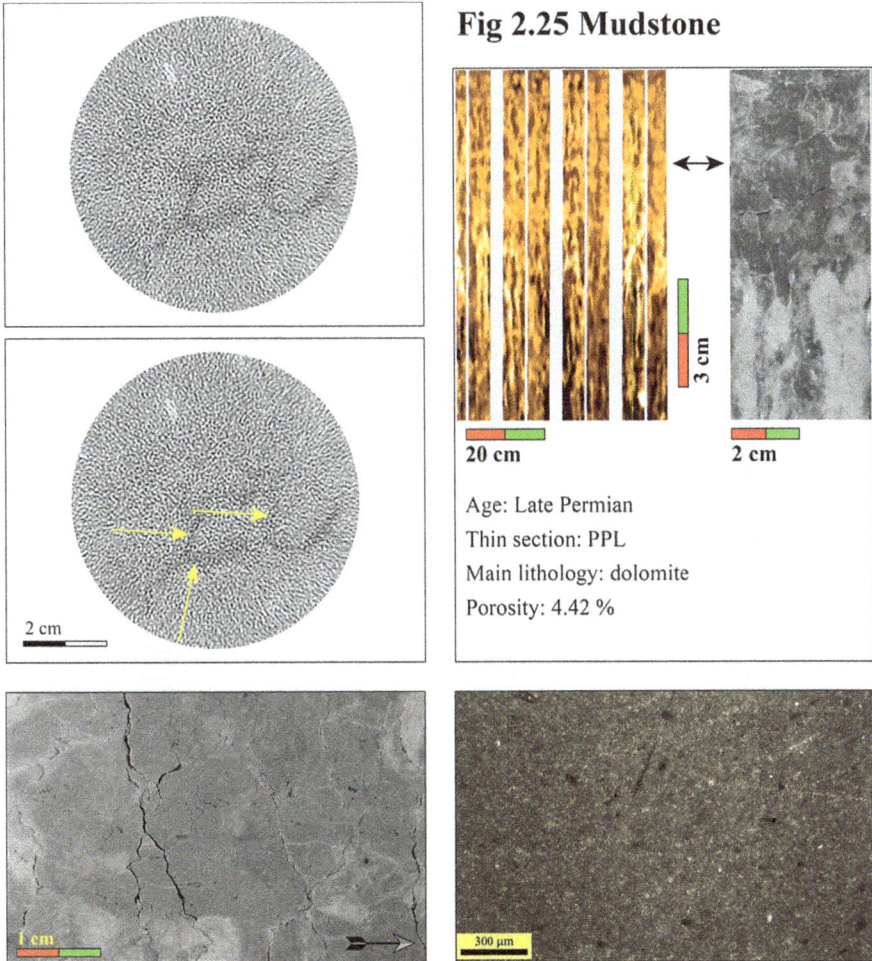

Fig 2.25 Mudstone

Age: Late Permian
Thin section: PPL
Main lithology: dolomite
Porosity: 4.42 %

FIGURE 2.25

A carbonate rock with mudstone texture has considerable amounts of micrite (allochems are less than 10 %). They are formed in low-energy environments and are nearly homogenous in CT images. Frequent thin stylolites in this sample are recognizable with darker color and hair-like pattern (arrows) in CT slices. High organic content can be inferred from dark color of thin section and core photos.

Fig 2.26 Fenestral dolomitic mudstone

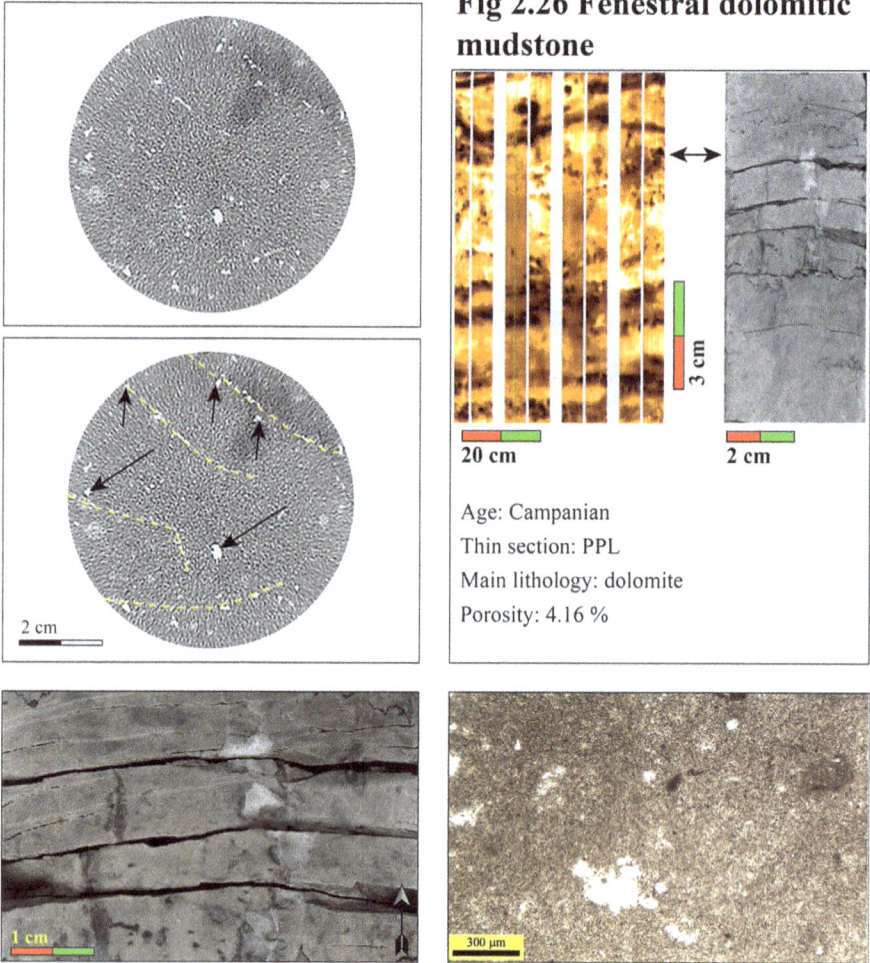

20 cm   2 cm

Age: Campanian
Thin section: PPL
Main lithology: dolomite
Porosity: 4.16 %

FIGURE 2.26

Fenestral porosities have been filled with anhydrite and so they are easily recognizable with white color in this sample (black arrows). These pore types are often arranged in parallel lines (dashed yellow lines) which are also parallel to the bedding planes. The sample is a laminated mudstone. Laminations are obvious in FMI, whole core and close-up photos.

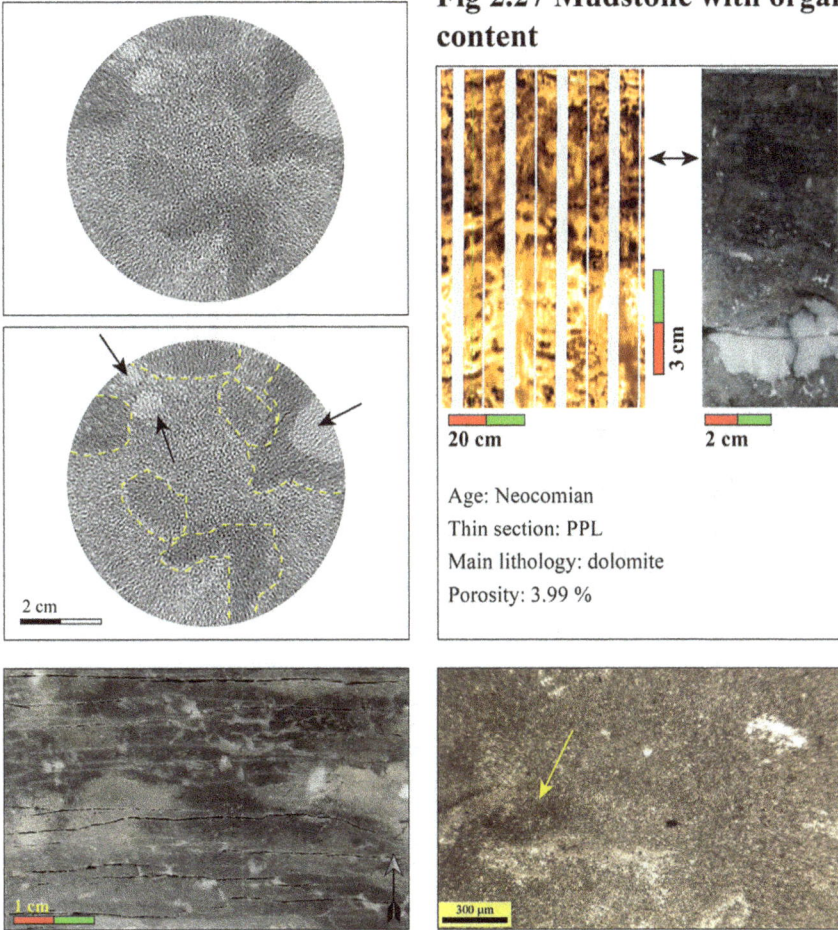

Fig 2.27 Mudstone with organic content

Age: Neocomian
Thin section: PPL
Main lithology: dolomite
Porosity: 3.99 %

FIGURE 2.27
Shallow water carbonate mudstones contain high organic content in many cases. These organic materials are marked by dark gray color and large and irregular shape in CT images (dashed yellow lines). Anhydrite nodules are also present in this sample (black arrows), which have been formed due to high rate of evaporation. Organic materials can also be seen in thin section studies (yellow arrow).

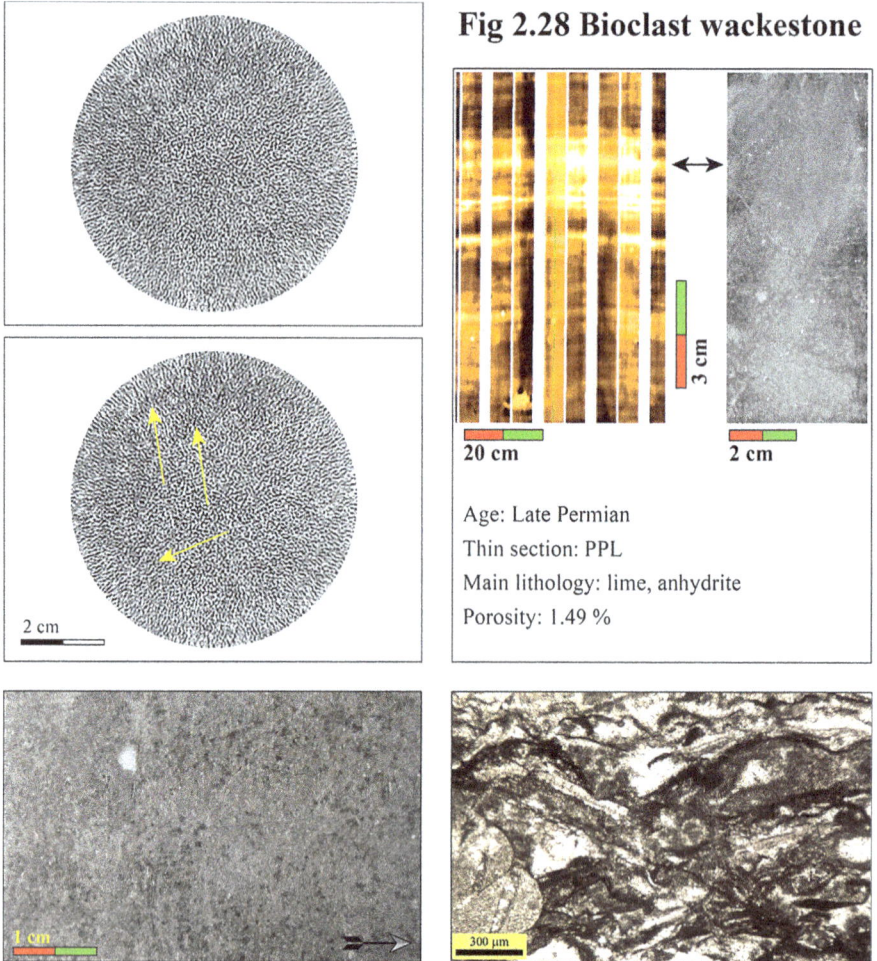

Fig 2.28 Bioclast wackestone

Age: Late Permian
Thin section: PPL
Main lithology: lime, anhydrite
Porosity: 1.49 %

FIGURE 2.28

Carbonate rocks without special features are nearly homogenous in CT images. Here, a wacke-stone texture is observed. Calcite, dolomite and anhydrite are present in the sample (thin sec-tion studies). Ghosts of some allochems are seen in the CT image with different shades of gray (yellow arrow).

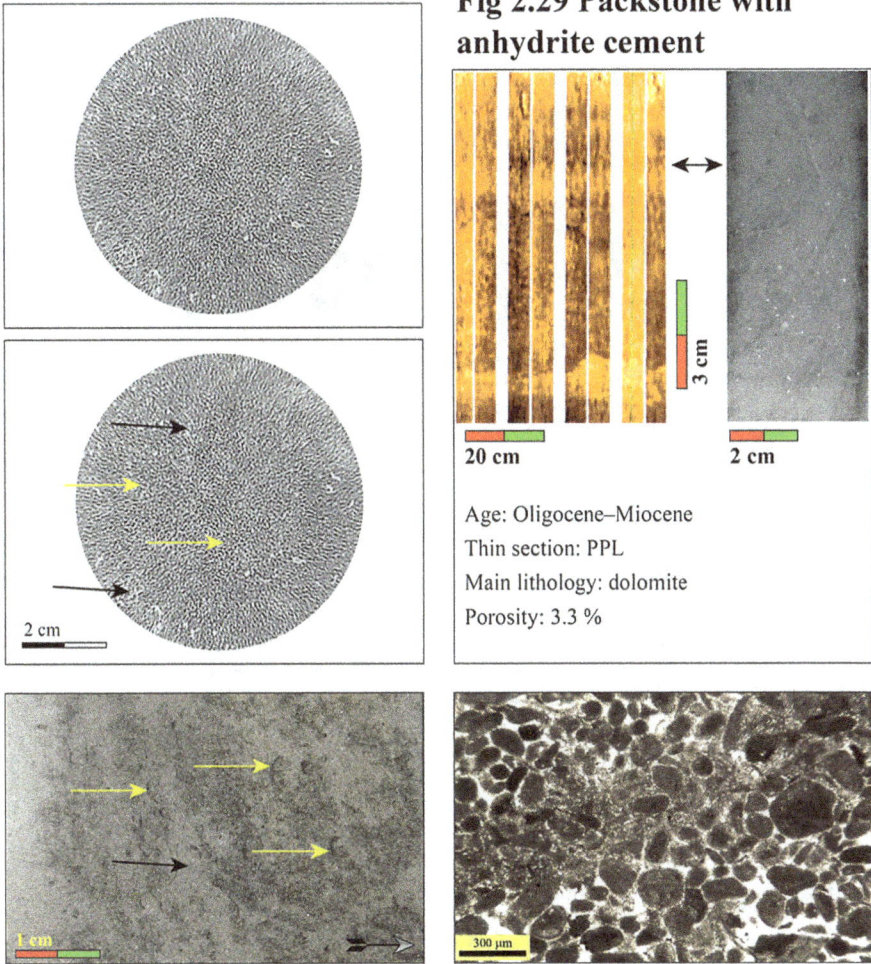

Fig 2.29 Packstone with anhydrite cement

Age: Oligocene–Miocene
Thin section: PPL
Main lithology: dolomite
Porosity: 3.3 %

FIGURE 2.29
Both grains and micrite matrix are present in this sample. Anhydrite cements also fill some pore spaces. Grains (yellow arrows) and anhydrite cements (black arrows) are clearly visible on core images. They are also recognizable on CT scan slices. The CT image has a salt–pepper pattern and a collection of light and dark pixels are visible. This is the same for the FMI image.

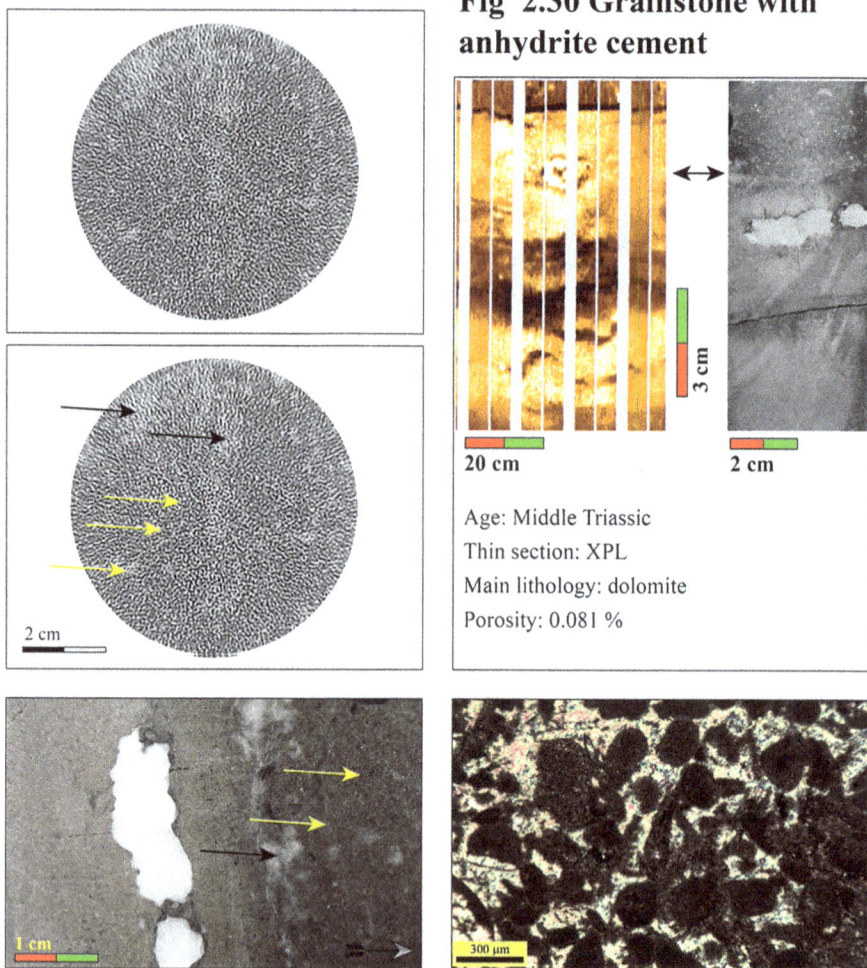

Fig 2.30 Grainstone with anhydrite cement

20 cm   2 cm

Age: Middle Triassic
Thin section: XPL
Main lithology: dolomite
Porosity: 0.081 %

FIGURE 2.30

A grainstone texture with anhydrite cement is obvious in these pictures. Grains (yellow arrows) are darker in CT images. Light anhydrites (black arrows) are visible between the grains or as separate nodules. Regarding the high frequency of grain and presence of evaporative minerals, the sample is related to a shallow, high-energy peritidal environment.

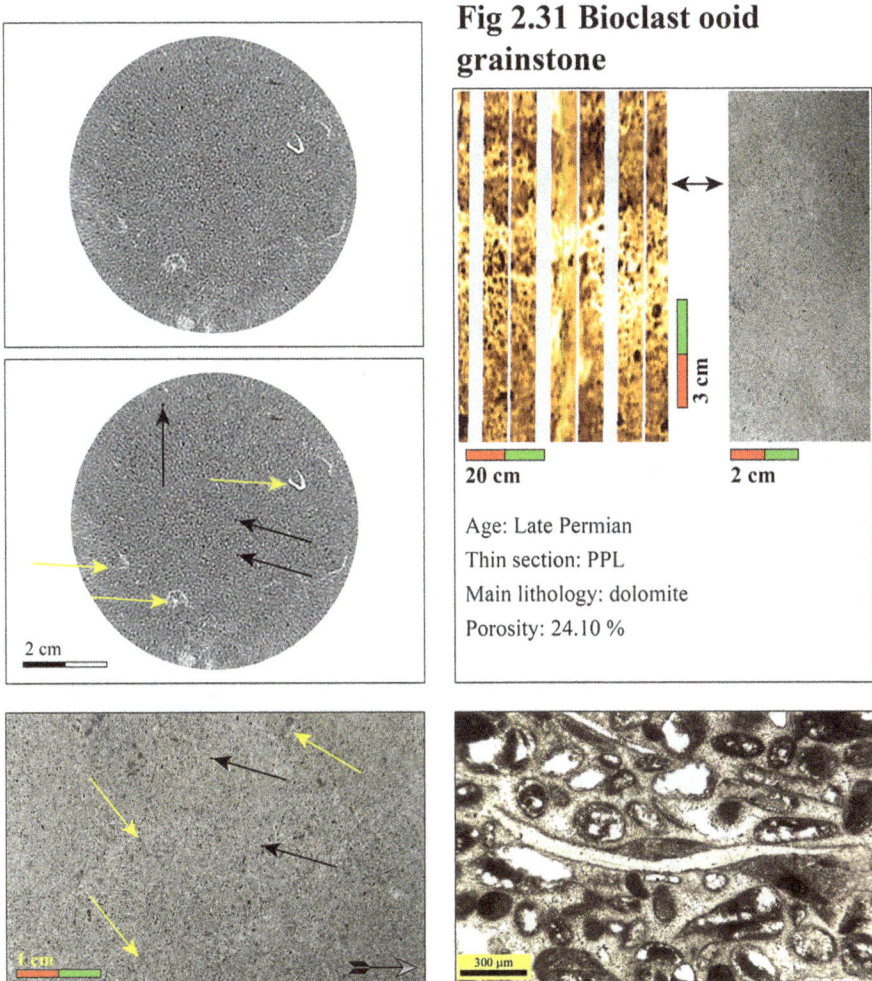

Fig 2.31 Bioclast ooid grainstone

Age: Late Permian
Thin section: PPL
Main lithology: dolomite
Porosity: 24.10 %

FIGURE 2.31

Features with different densities can be recognized in CT images. Bioclasts have been filled with calcite crystals while ooids have been dissolved in this sample. So, bioclasts are denser and lighter than background (yellow arrows) and ooids are seen as black spots (black arrows). The sample is grainstone and so background of the image is homogenous.

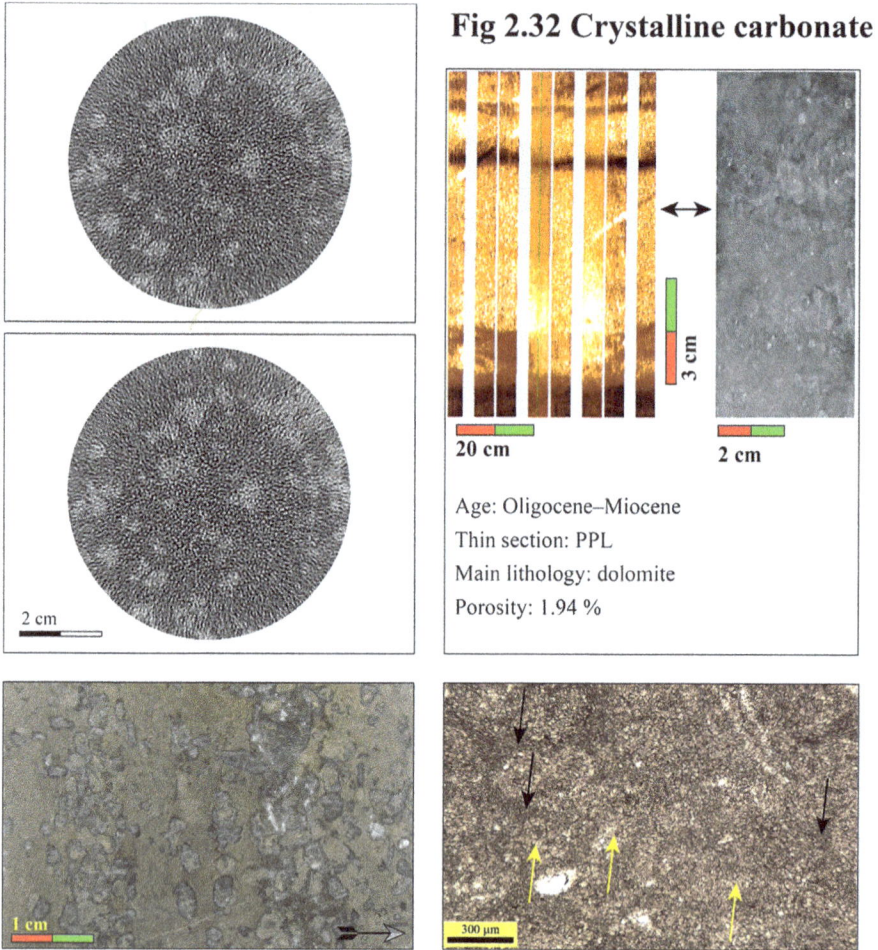

Fig 2.32 Crystalline carbonate

Age: Oligocene–Miocene
Thin section: PPL
Main lithology: dolomite
Porosity: 1.94 %

FIGURE 2.32
The name "crystalline carbonate" is used for the carbonate samples when the primary fabric is not recognizable. It is believed that dolomitization increases the porosity of the rocks. This sample has salt–pepper view in CT images because two dolomite generations are present. Fine dolomites are marked by dark gray (higher porosity, black arrows) while large crystals are lighter (yellow arrows).

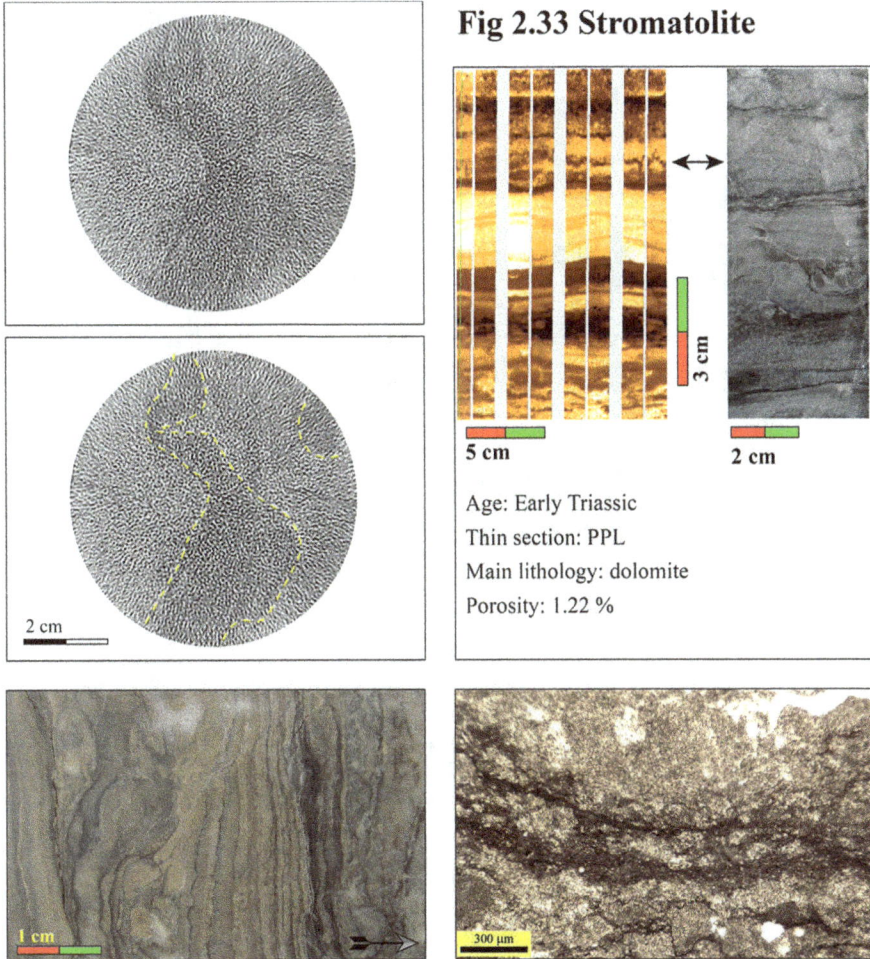

Fig 2.33 Stromatolite

Age: Early Triassic
Thin section: PPL
Main lithology: dolomite
Porosity: 1.22 %

FIGURE 2.33

The cross-section of two planes is a line and so layered features are marked as lines in CT images. The thickness of this line depends on the thickness of the layer. Here, a stromatolite layer with high organic content is seen (dashed yellow lines). The alternation of light and dark gray color is also visible in whole core, FMI, thin section and close-up photos.

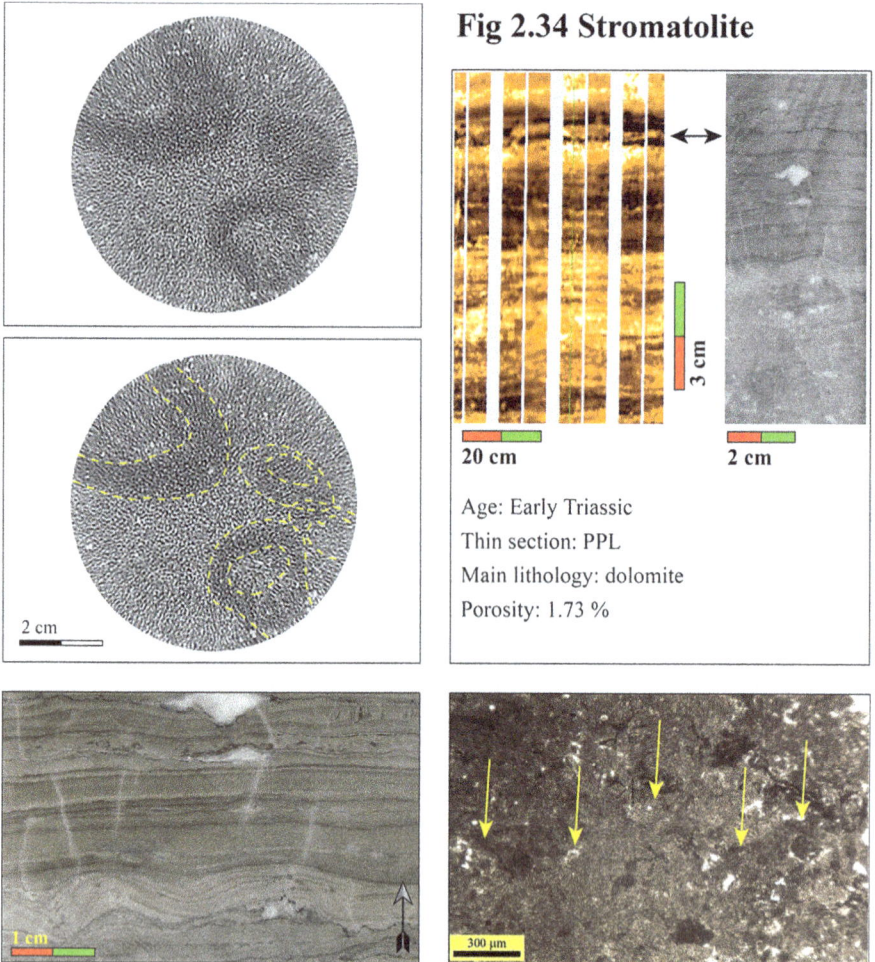

Fig 2.34 Stromatolite

Age: Early Triassic
Thin section: PPL
Main lithology: dolomite
Porosity: 1.73 %

FIGURE 2.34

Stromatolites are layered carbonate rocks which are created by photosynthetic microorganisms, mainly cyanobacteria. They are formed as very fine layers and so can be distinguished by alternations of dark and white bands in CT images (dashed yellow lines). They are also visible in FMI, whole core and close-up photos. These structures are also recognizable in thin sections (a curved surface with is marked by yellow arrows).

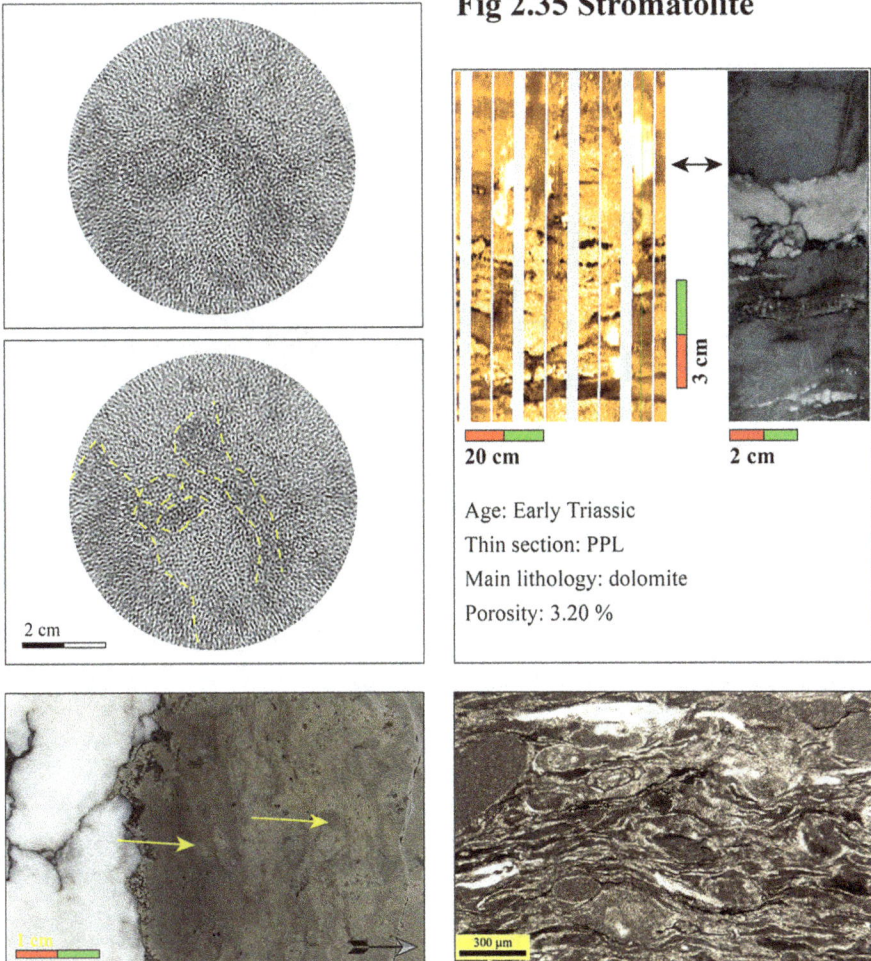

Fig 2.35 Stromatolite

Age: Early Triassic
Thin section: PPL
Main lithology: dolomite
Porosity: 3.20 %

FIGURE 2.35

Stromatolites are microbial sedimentary layers which are routinely lived in shallow environments, such as peritidal settings. They are recognized by their layered structure in CT scan images. Layers are wavy in most cases and some dome-like features of microorganisms are also present (arrows on close-up photo). Therefore, the cross-section is a closed area or curved line in CT slices (dashed lines). These layers contain high organic content and so are darker in color.

Fig 2.36 Thrombolite

20 cm   2 cm

Age: Early Triassic
Thin section: PPL
Main lithology: lime
Porosity: 1.34 %

FIGURE 2.36

Clotted fabrics of thrombolites are formed by cyanobacteria in carbonate formations. They are observed at the scale of millimeters to centimeters. These structures are routinely formed at the upper part of shoreline. At the large mass extinctions of the Earth, they bloom and form large colonies throughout the world. This sample contains thrombolites from the base of Triassic post extinction facies. They are recognized by dark color in CT images due to high organic content.

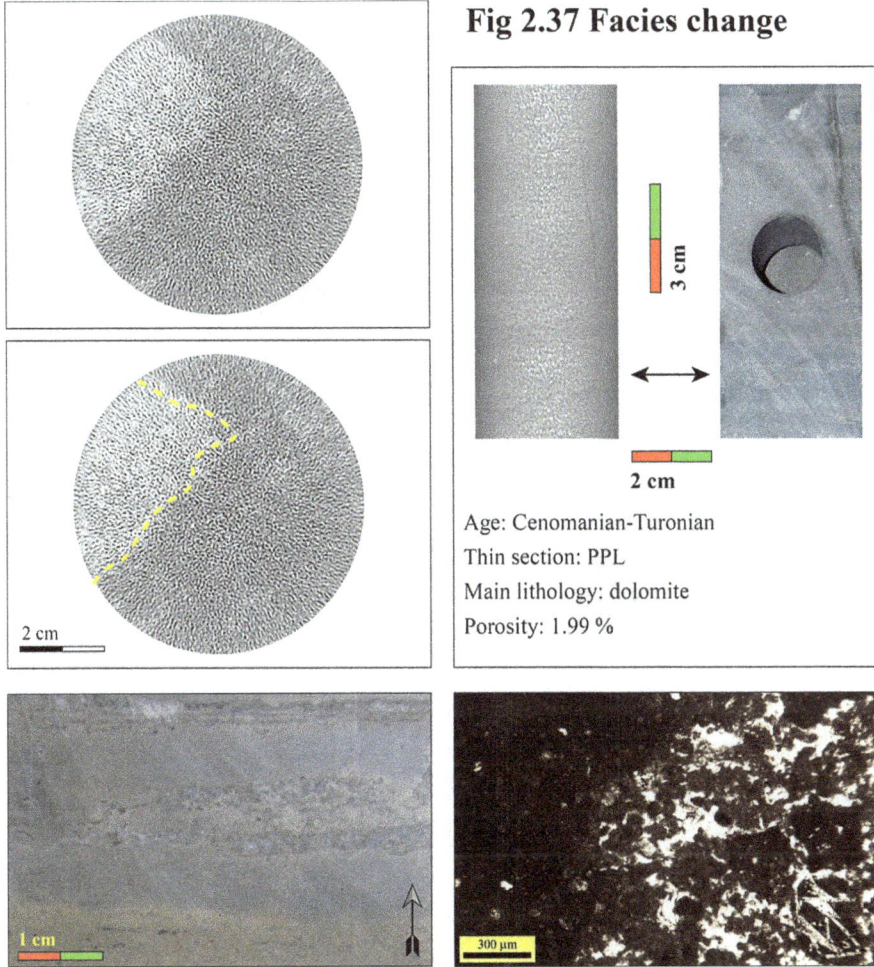

Fig 2.37 Facies change

Age: Cenomanian-Turonian
Thin section: PPL
Main lithology: dolomite
Porosity: 1.99 %

FIGURE 2.37

Changes in reservoir rock properties in a vertical column show the variations in physical, chemical and biological conditions of the sedimentary environments, or later diagenetic changes. These changes can be seen on CT images. Here, change from one facies (mudstone, darker part) to another (anhydrite cemented intraclast packstone, lighter part) is seen in the image. The lower intensity in gray color is due to the more anhydrite content in the packstone facies.

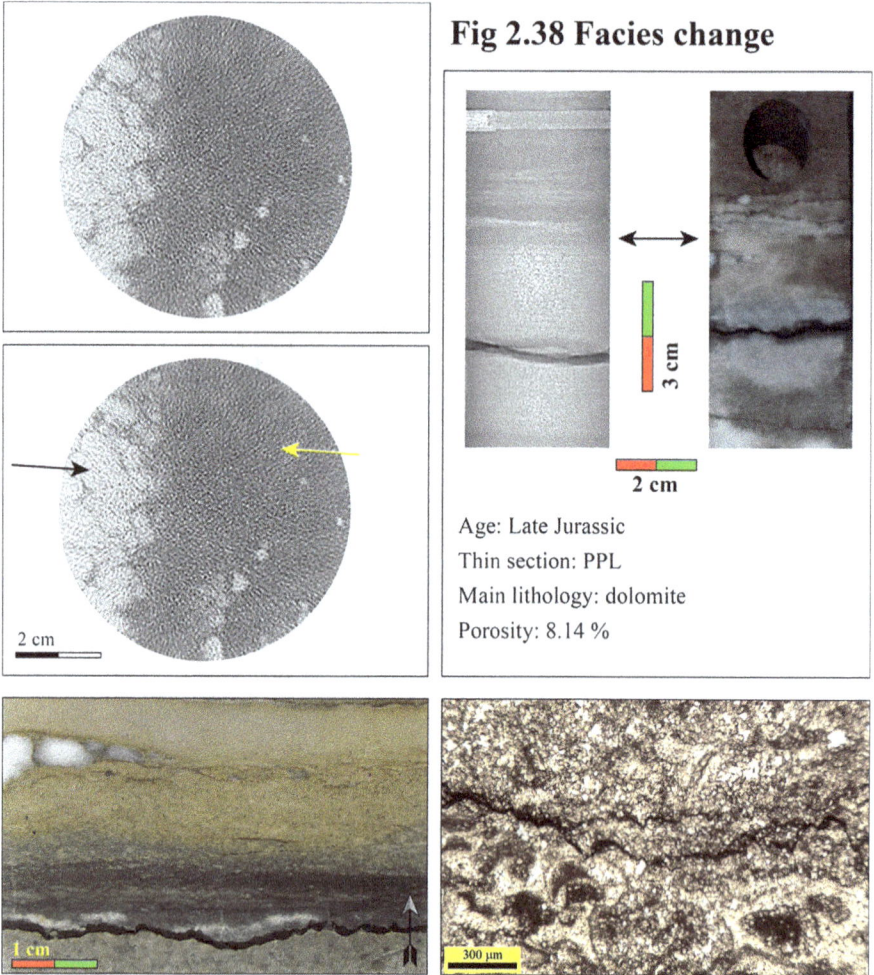

Fig 2.38 Facies change

2 cm

Age: Late Jurassic
Thin section: PPL
Main lithology: dolomite
Porosity: 8.14 %

2 cm

FIGURE 2.38

Heterogeneity is an intrinsic property of carbonate reservoirs. This includes variations in environmental conditions at the time of deposition, which is reflected in the facies characteristics. Minerals can be differentiated on CT images based on the changes in their density. Here, the presence of more anhydrite, with higher density, caused lighter color (milky white) in the left part of the transverse CT image. This change in facies is also detectable in longitudinal CT image, whole and close up core photos, as well as thin sections.

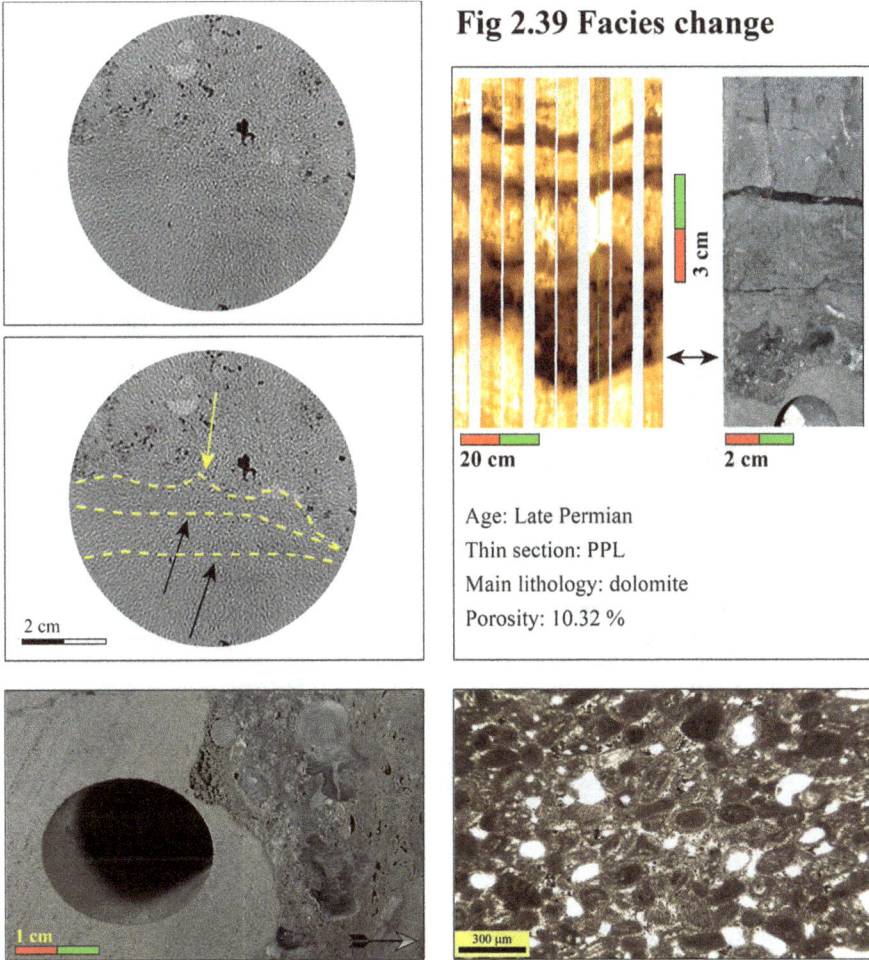

Fig 2.39 Facies change

Age: Late Permian

Thin section: PPL

Main lithology: dolomite

Porosity: 10.32 %

FIGURE 2.39

Facies changes from packstone to grainstone with an angular score surface (yellow arrow) and so it can be viewed on CT images. It is clearly seen on FMI image, too. Lamination is obvious in grainstone texture (black arrow). The packstone has messy texture because of the change in the depositional energy. Pore spaces are present in both textures, but they have more regular shapes in grainstone facies.

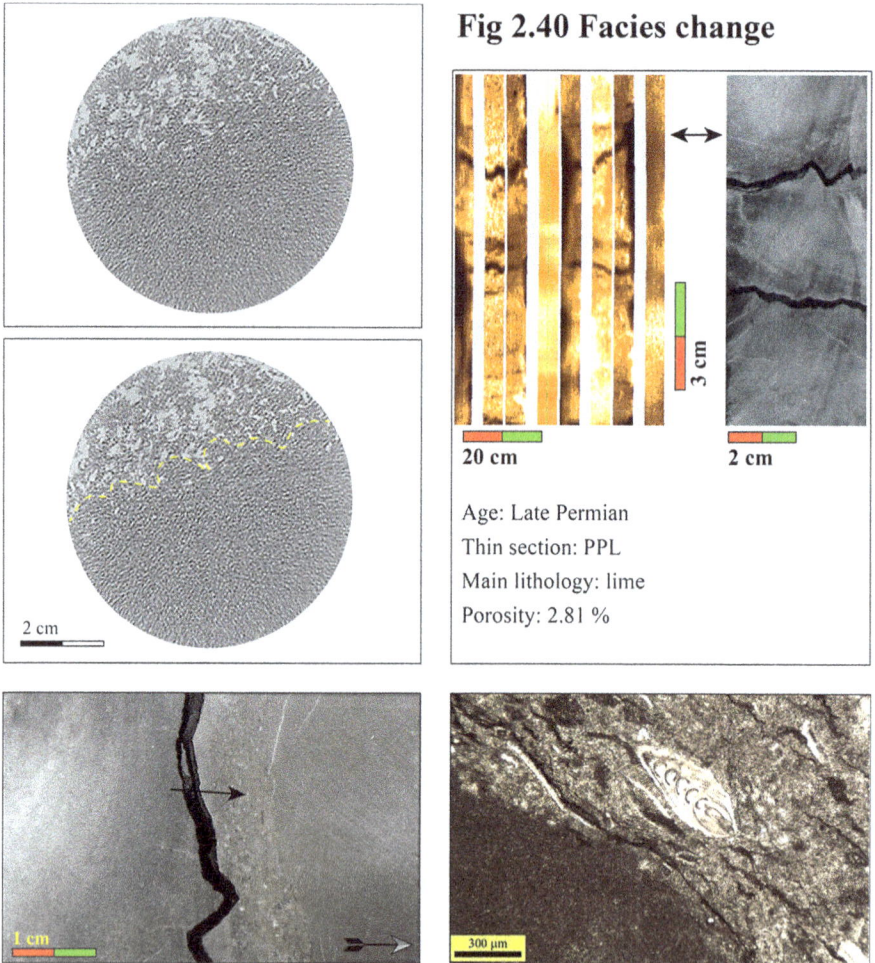

Fig 2.40 Facies change

Age: Late Permian
Thin section: PPL
Main lithology: lime
Porosity: 2.81 %

FIGURE 2.40

Change in the texture and facies of carbonates indicates variations in physical, chemical or biological conditions in sedimentary environment. Here, a layer of grain-dominated texture (black arrow) has been deposited between the mudstones. Grains have been filled with calcite cements and therefore have higher density than micrite or microspar background. Different textures (and so densities) have been traced with different gray intensities in CT image (yellow dashed line).

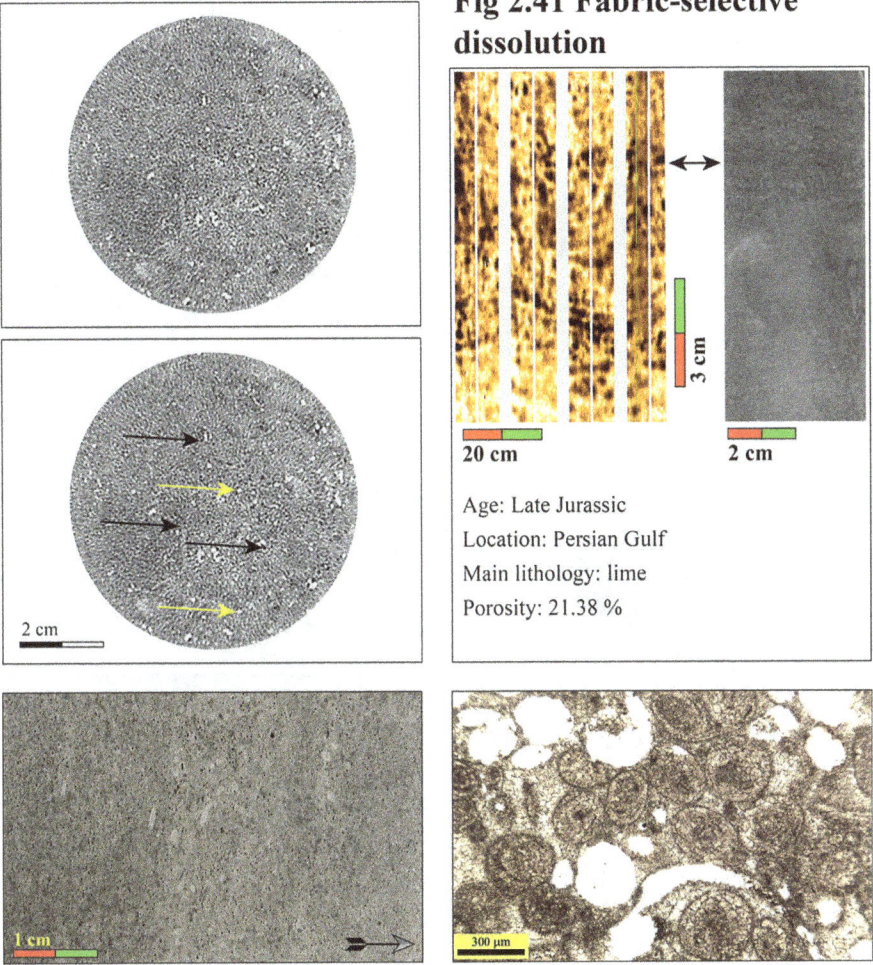

Fig 2.41 Fabric-selective dissolution

Age: Late Jurassic
Location: Persian Gulf
Main lithology: lime
Porosity: 21.38 %

FIGURE 2.41

Dissolution could be fabric-selective or fabric-destructive. Ooids and bioclasts have been dissolved in this sample. They are recognizable by dark color (black arrows) as compared to the limy background. Rhombic dolomite crystals are clearly visible (yellow arrows). Dolomites are lighter than the background due to their higher density.

Pore types have strong effects on carbonate reservoir behavior. Therefore, this chapter also considers the trace of various pore types in CT images (Figures 2.41–2.51). Moldic and vuggy pores are seen in these images in many cases. They are characterized with dark color due to low density (fluid or empty spaces). In addition to these routine features, there are also some other facies or diagenetic properties that can be recognized in CT photos.

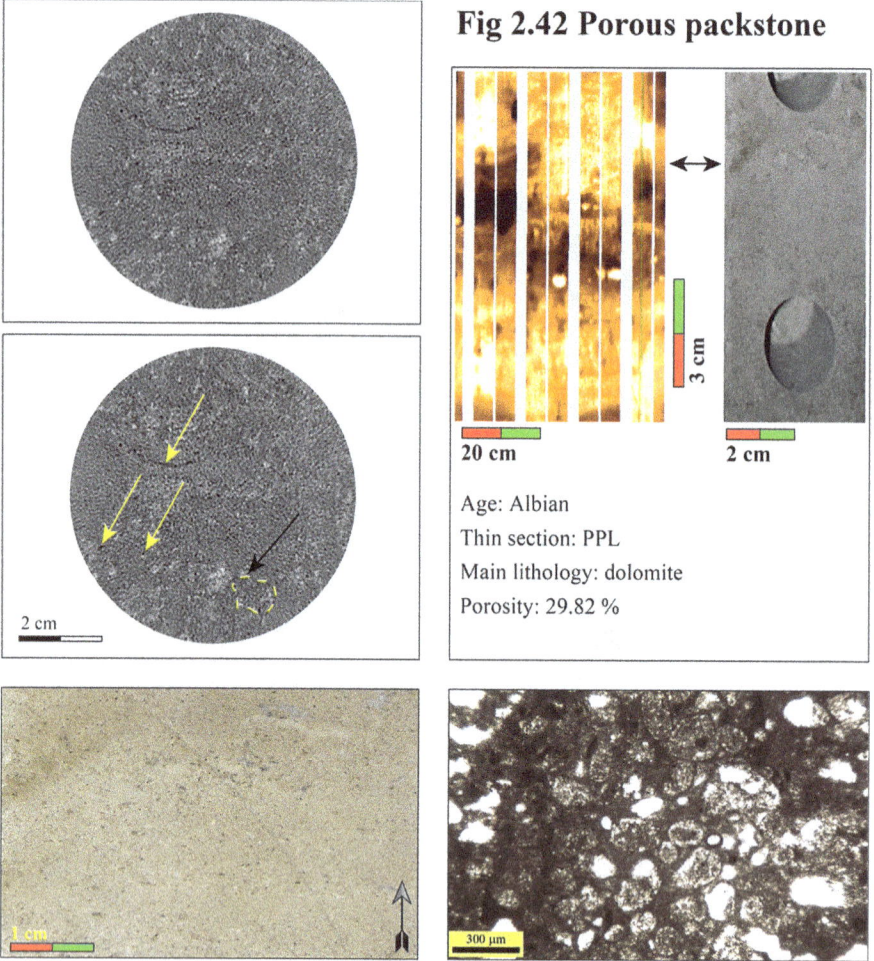

Fig 2.42 Porous packstone

Age: Albian
Thin section: PPL
Main lithology: dolomite
Porosity: 29.82 %

FIGURE 2.42
This packstone sample contains high frequency of moldic pores (yellow arrows). The remaining grains have been dolomitized and so the CT image is a mixture of light and dark spots. The background is gray, which shows the mud matrix. Micrite envelope of some grains can be distinguished with dark lines (black arrow).

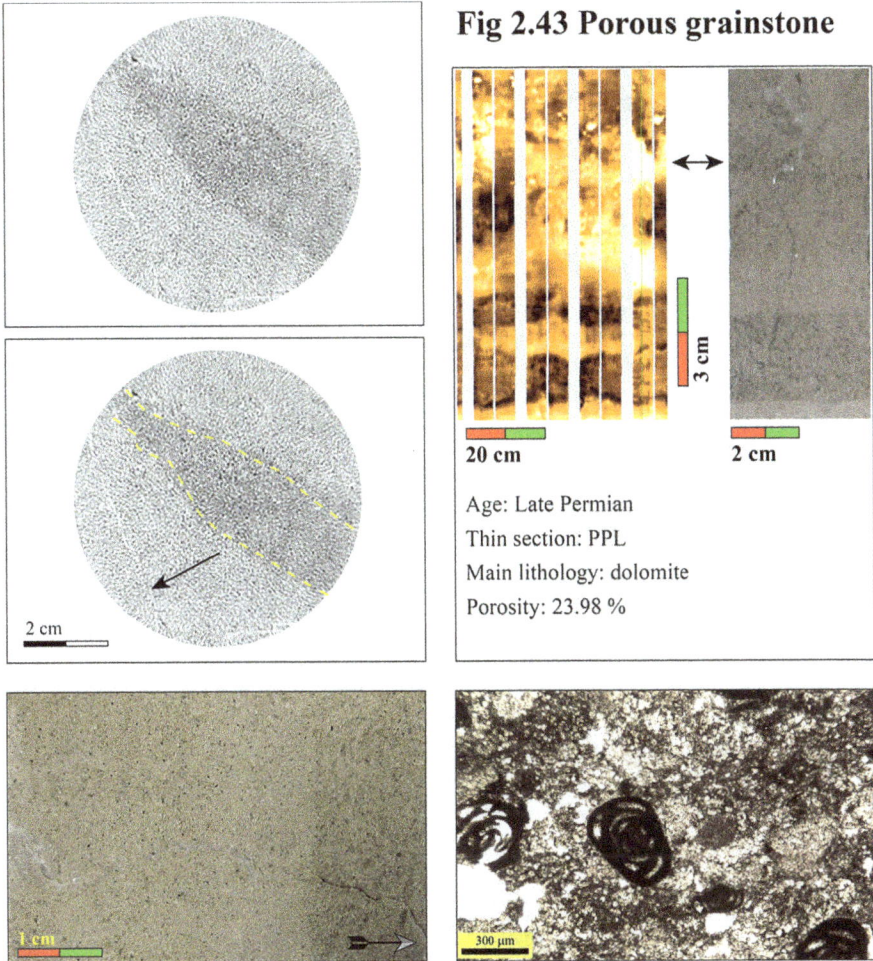

Fig 2.43 Porous grainstone

Age: Late Permian
Thin section: PPL
Main lithology: dolomite
Porosity: 23.98 %

FIGURE 2.43

Grainstones are more homogenous as compared to wackestones and packstones and this is reflected in their CT images. Here, a layered porous grainstone is visible. Moldic pores are more frequent in some layers. One of these layers can be observed on CT image as a darker part (dashed yellow lines). A vertical filled fracture is recognizable as a light line (arrow).

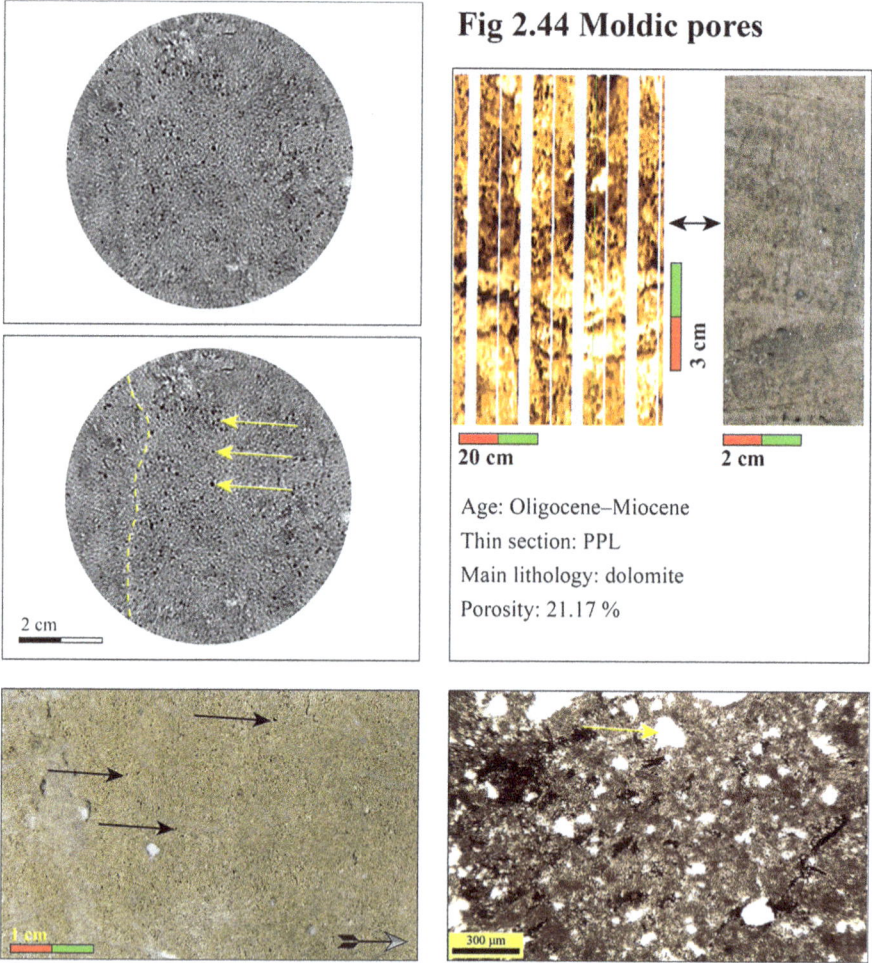

Fig 2.44 Moldic pores

Age: Oligocene–Miocene

Thin section: PPL

Main lithology: dolomite

Porosity: 21.17 %

FIGURE 2.44

Moldic pores are easily recognizable on CT images with their dark color and specific shape. These images show an ooid grainstone sample with oomoldic pores. The pores are visible on FMI, core and thin section photos. The FMI image shows a dotty pattern. Arrows show the moldic porosities in all cases. Cross-laminations are evident in whole core, FMI and CT images (yellow dashed line).

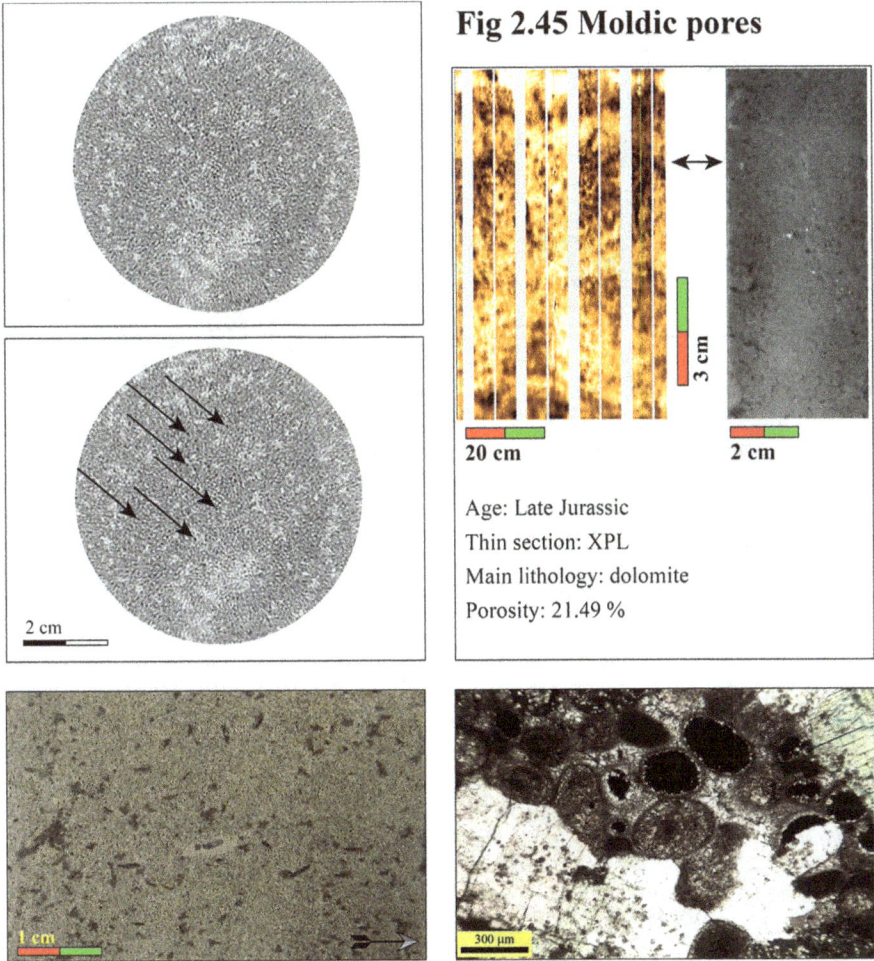

Fig 2.45 Moldic pores

Age: Late Jurassic

Thin section: XPL

Main lithology: dolomite

Porosity: 21.49 %

FIGURE 2.45

Very frequent moldic pores are visible in this sample. The pores are recognized by their black color in CT images (arrows). The sample is composed of lime, dolomite and anhydrite and therefore a salt–pepper texture has been formed. Empty spaces are visible on close-up, thin section and FMI images. Anhydrite is frequent in the sample, which is marked by its white color in CT slice.

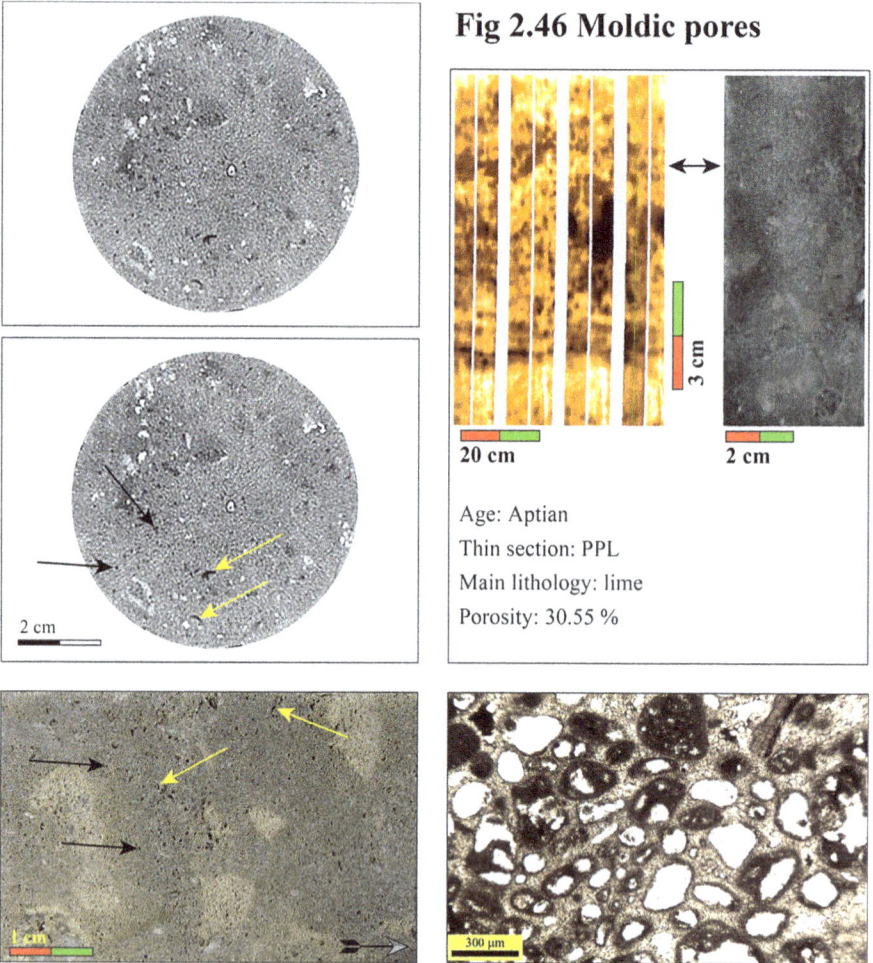

Fig 2.46 Moldic pores

20 cm   2 cm

Age: Aptian
Thin section: PPL
Main lithology: lime
Porosity: 30.55 %

FIGURE 2.46

Moldic pores are routinely created in meteoric diagenetic environment. Unstable carbonate minerals such as aragonite and low magnesium calcite are prone to dissolution. Here, the mold of both bioclasts (yellow arrows) and ooids (black arrows) can be seen. Intraclasts are made of micrites and so can be recognized by their darker color compared to calcite cements.

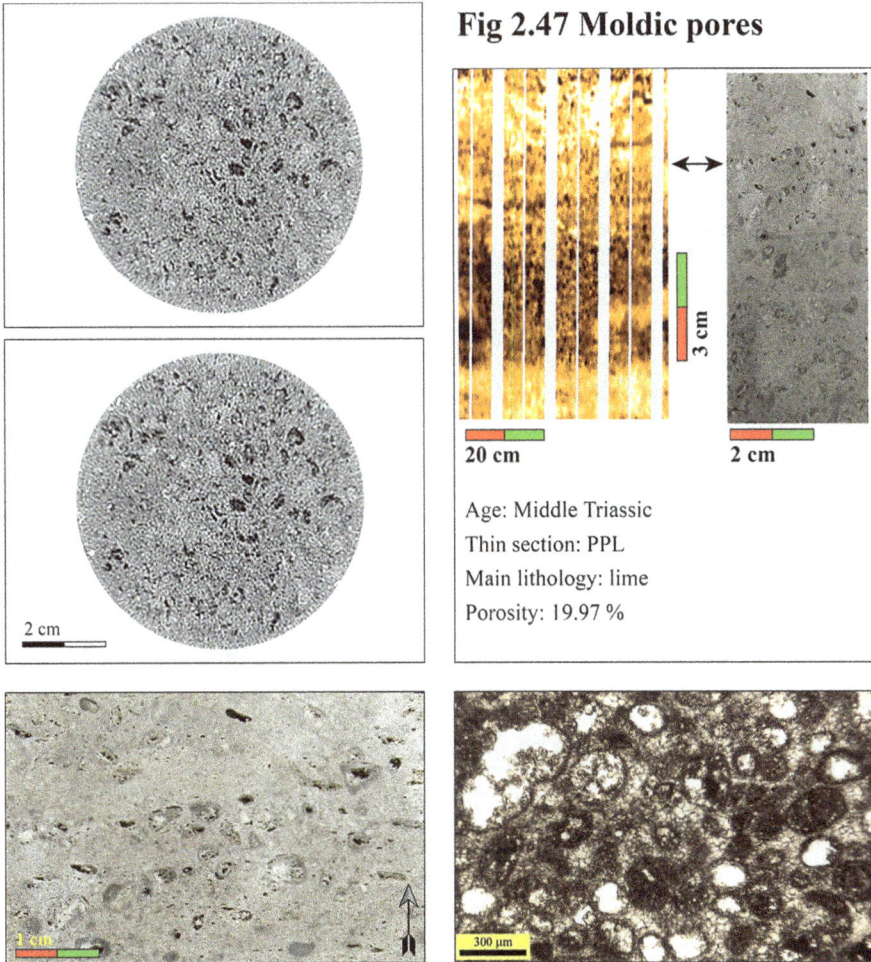

Fig 2.47 Moldic pores

Age: Middle Triassic

Thin section: PPL

Main lithology: lime

Porosity: 19.97 %

20 cm

2 cm

2 cm

300 µm

FIGURE 2.47

Large bioclasts have been dissolved and their molds are recognizable by black color in CT images. These pores are not connected to each other. The sample porosity is 19.97 % while permeability is 0.64 mD. Both FMI and core photos also show the molds of these bioclasts. Meteoric waters in meteoric diagenetic environment are responsible for this kind of dissolution in carbonate reservoirs.

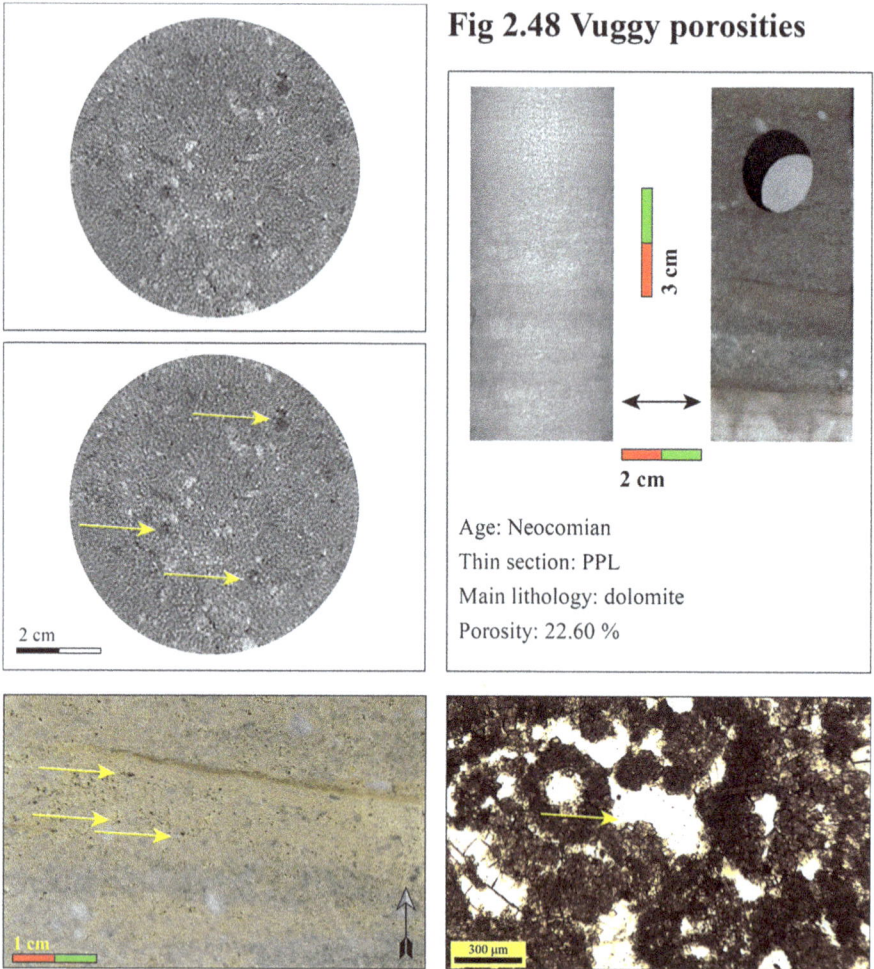

Fig 2.48 Vuggy porosities

Age: Neocomian
Thin section: PPL
Main lithology: dolomite
Porosity: 22.60 %

FIGURE 2.48

Empty spaces are characterized with darker color (black) on CT scan images. Large vuggy pore spaces are easily recognizable on these images and also core photos. In this case, anhydrite nodules are also present in dolomitic matrix. Frequent allochems in high-energy environment have created low angle cross laminations, which are evident on whole and close up core photos as well as longitudinal CT image. They are not visible on transverse CT because they are nearly horizontal.

Fig 2.49 Vuggy pores

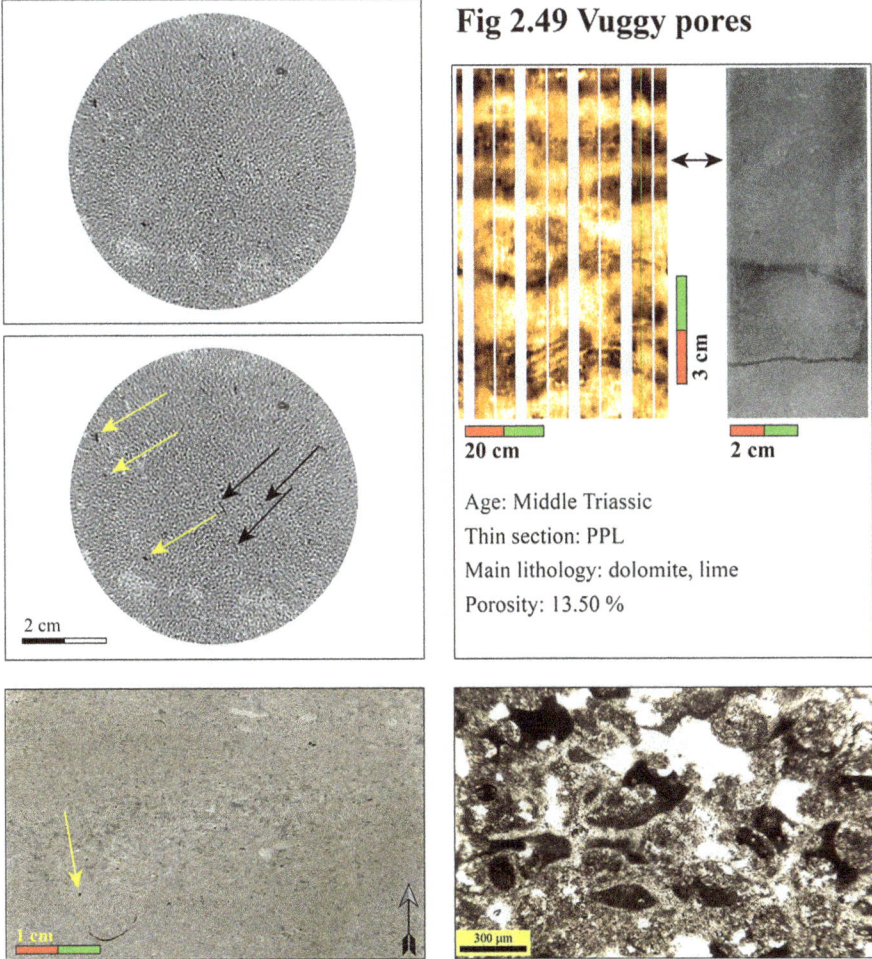

Age: Middle Triassic

Thin section: PPL

Main lithology: dolomite, lime

Porosity: 13.50 %

FIGURE 2.49

Vuggy pores are larger than grains and so they are easily recognized on CT images by their black color. In this sample, frequent vuggy pores are oriented along a layer (black arrows). There are also some other pores which are scattered in the sample (yellow arrows). White patches are dolomitic parts. The pores are also visible in thin section studies.

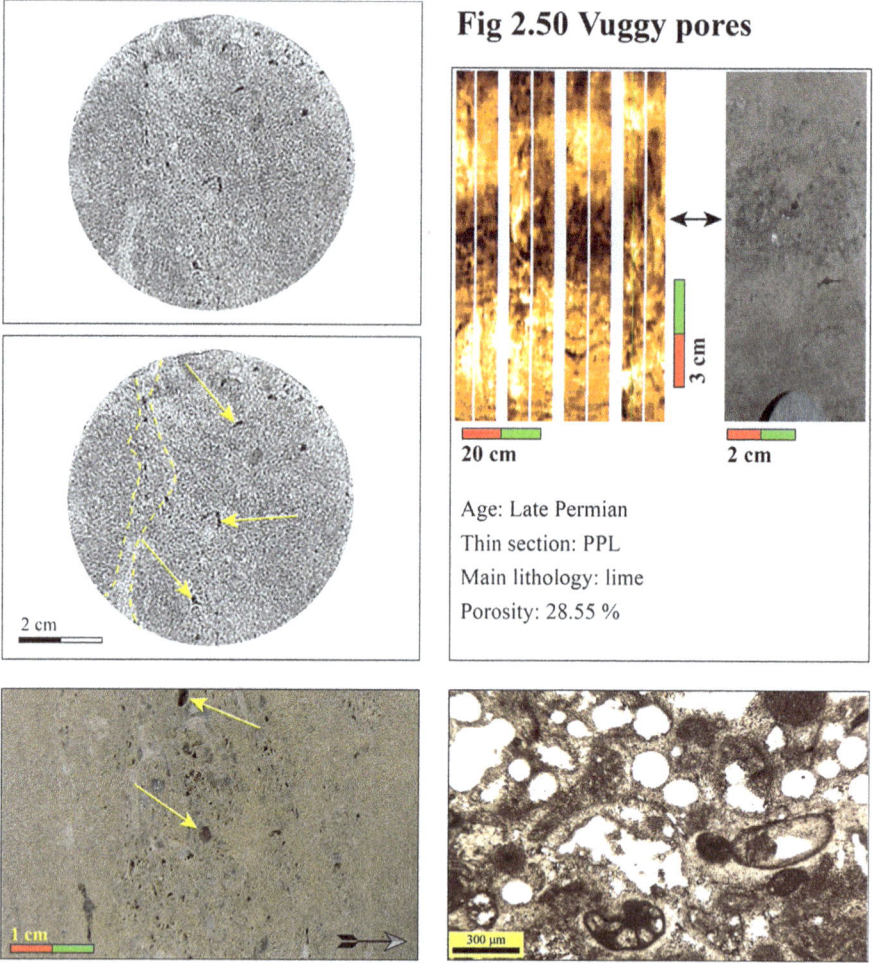

Fig 2.50 Vuggy pores

Age: Late Permian

Thin section: PPL

Main lithology: lime

Porosity: 28.55 %

FIGURE 2.50

Vuggy pores are obviously seen in these pictures. They are marked by dark gray to black color in CT images (yellow arrows). Vuggy porosity is routinely created by enlargement of a precursor moldic pore. Therefore, they are more frequent in grain-dominated layers. Layering, which has been created at the time of deposition (traced layer on CT image), is also obvious in this sample. The porosity of the sample is high.

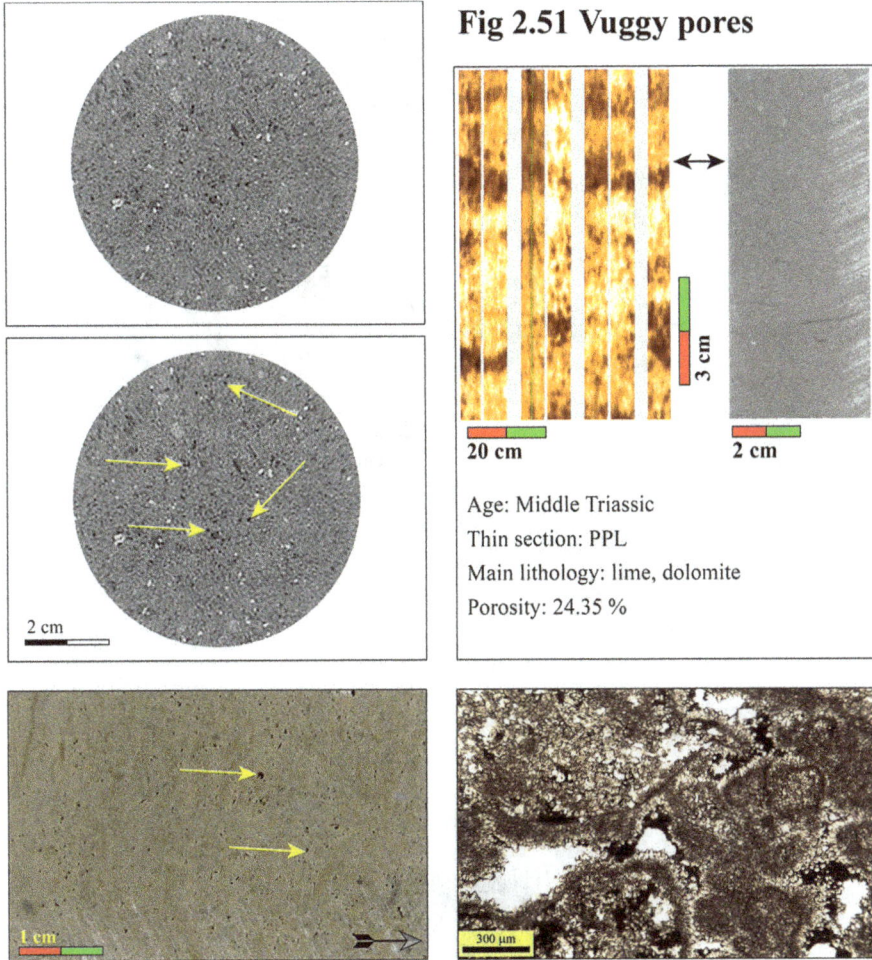

Fig 2.51 Vuggy pores

Age: Middle Triassic

Thin section: PPL

Main lithology: lime, dolomite

Porosity: 24.35 %

FIGURE 2.51

Large pores are recognized by the black color in CT images (yellow arrows). Vugs have been developed in this dolomitic limestone sample due to dissolution. Anhydrites are marked by white color (about 10 %) in CT images. Vugs are recognizable in FMI log by dark color due to low resistivity. These pores are obvious on whole core and close-up photos too.

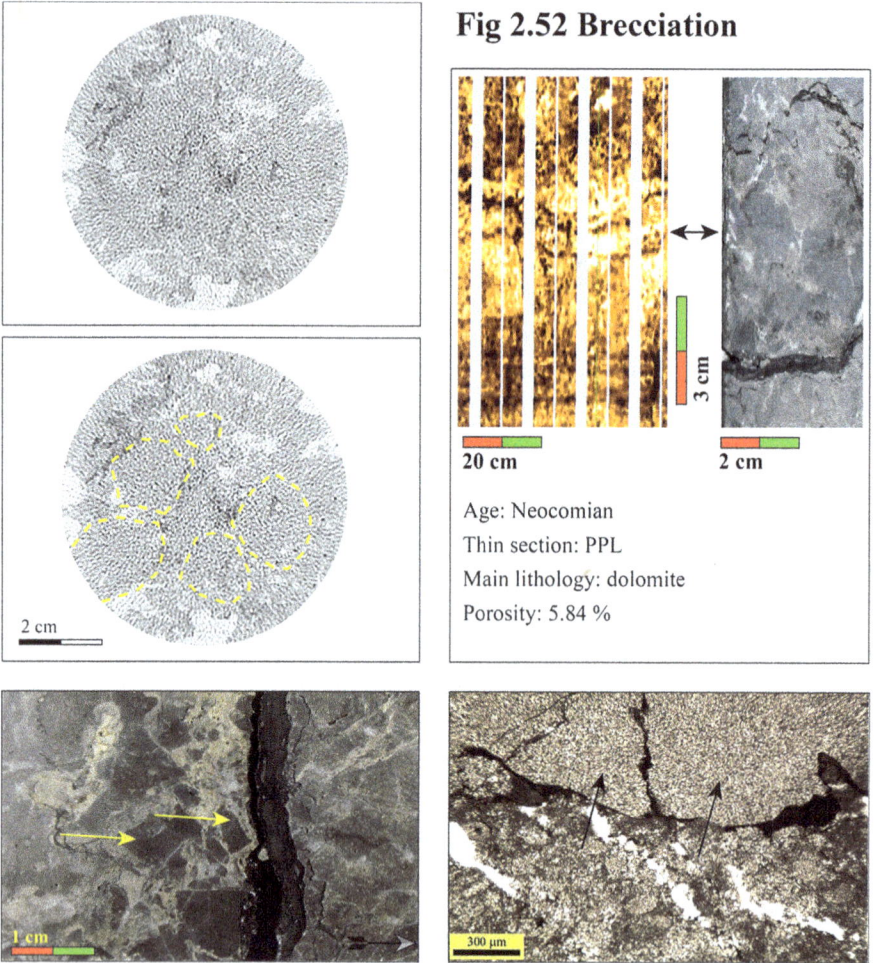

Fig 2.52 Brecciation

Age: Neocomian

Thin section: PPL

Main lithology: dolomite

Porosity: 5.84 %

FIGURE 2.52

Brecciation is occurred in carbonate-evaporite series due to the difference in rock density. It can be distinguished with closed areas in CT images with different gray intensity. Matrix and clasts have the same lithology (dolomite in this case) and so similar gray color in CT slices, but the intergrain space is filled with high-density materials such as anhydrite or iron rich thin layers. Arrows and dashed lines show the clasts.

Brecciation is frequently seen in carbonate environments. This is due to the difference of mineral density (calcite, dolomite and anhydrite), dissolution or high porosity of sediments at the time of deposition (Figures 2.52–2.55).

Some examples of geological features which are presented in this chapter include chicken-wire fabric of anhydrites, agitation in the sedimentary environment and its effect on CT images, load structures, microbial encrustations

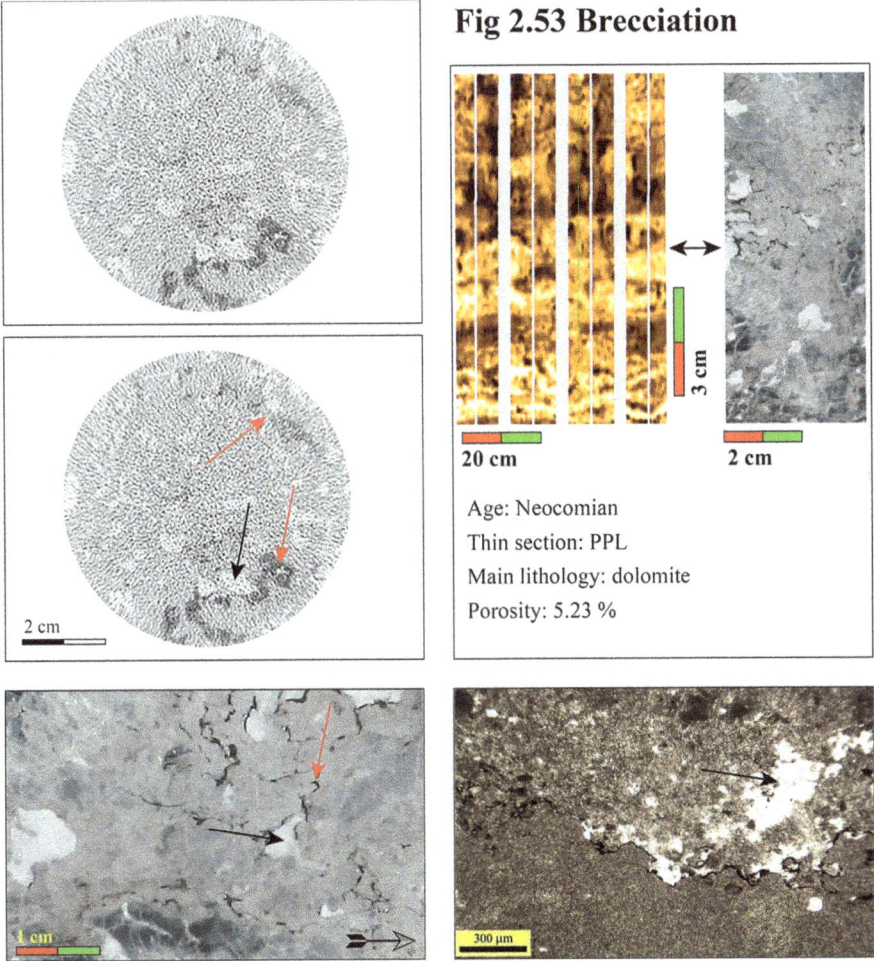

Fig 2.53 Brecciation

Age: Neocomian

Thin section: PPL

Main lithology: dolomite

Porosity: 5.23 %

FIGURE 2.53

The brecciation process is related to depositional environment or early diagenetic stage but clasts may be compacted in burial realm. In such cases, stylolites are formed at the boundary of the clasts. Both mechanical and chemical compactions are obvious in these images. The inter-grain spaces have been filled with anhydrite cement (black arrow) or low-density clay content (red arrow).

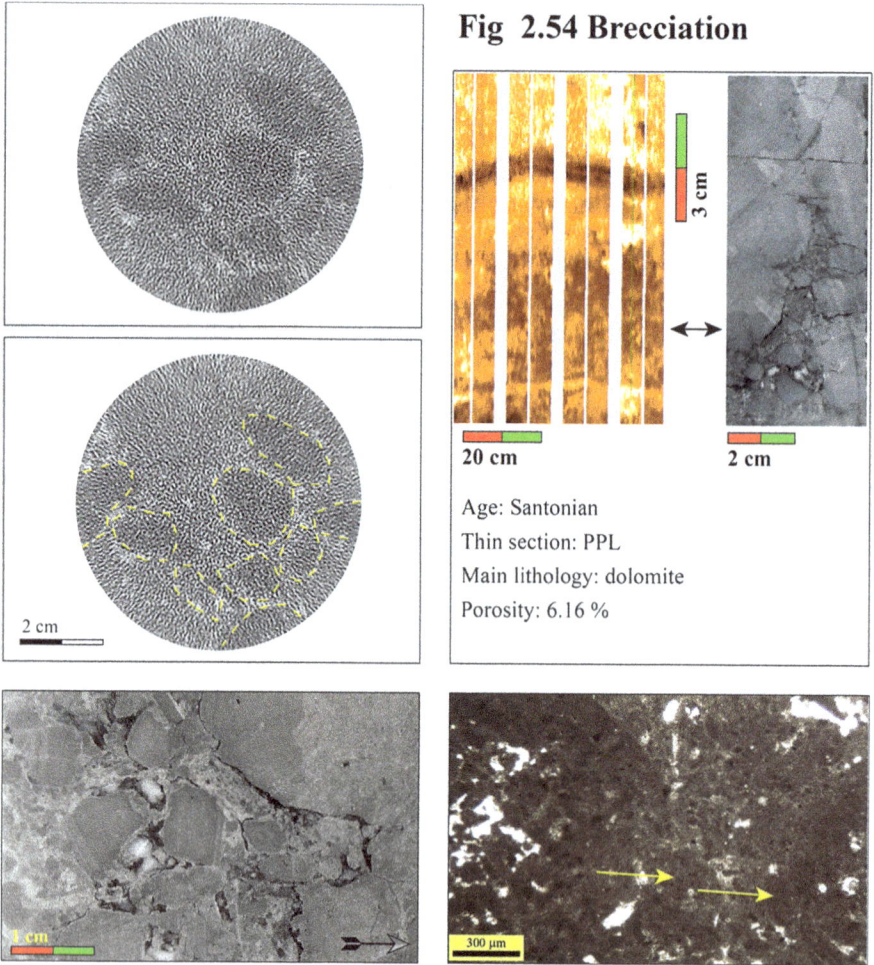

Fig 2.54 Brecciation

Age: Santonian

Thin section: PPL

Main lithology: dolomite

Porosity: 6.16 %

FIGURE 2.54

In carbonate depositional environments, brecciation happens due to the difference in the density of minerals, especially when anhydrite is present. Minor amounts of anhydrite (about 10 %) in this sample caused sediment collapse. Clasts are visible on CT images (yellow perimeters), thin section (arrows) and core photos. Clasts are angular which indicate in situ deposition.

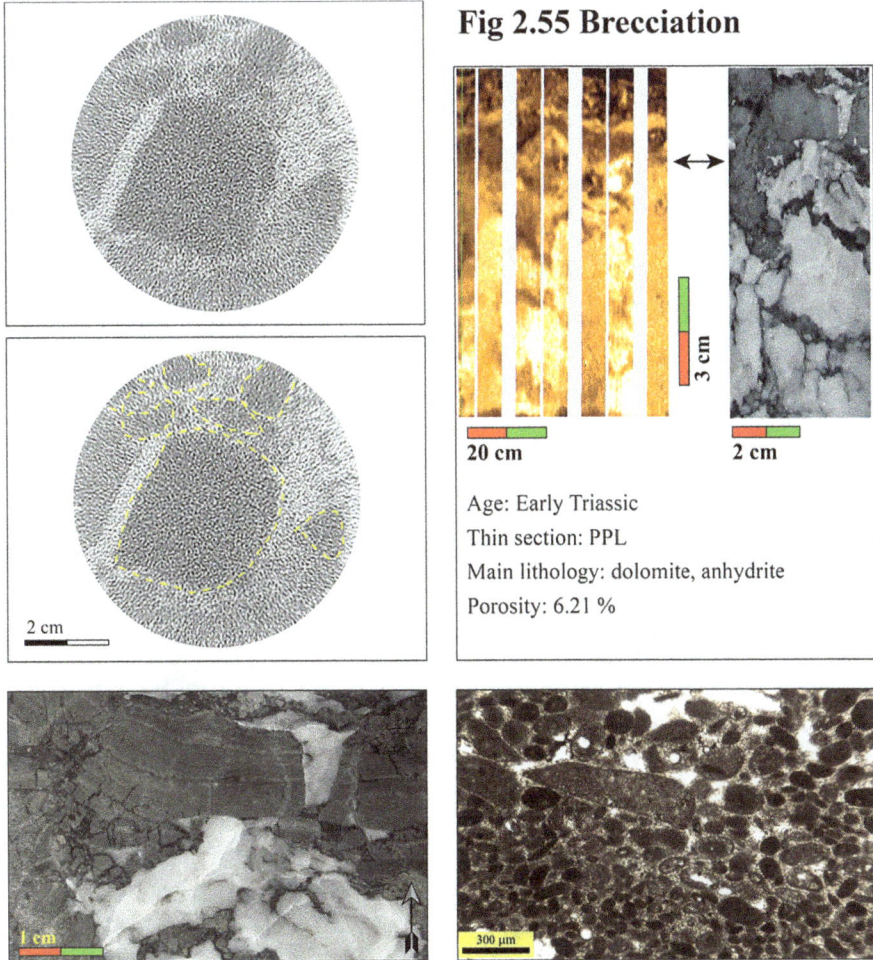

Fig 2.55 Brecciation

Age: Early Triassic

Thin section: PPL

Main lithology: dolomite, anhydrite

Porosity: 6.21 %

FIGURE 2.55

Brecciation take place in peritidal carbonate-evaporite environments due to the difference in sediment densities. Recognition of angular clasts is not easy because they have similar lithology and texture with the surrounding rocks. Anhydrite cements fill the intergrain spaces in some cases, which facilitate their recognition (dashed yellow perimeters).

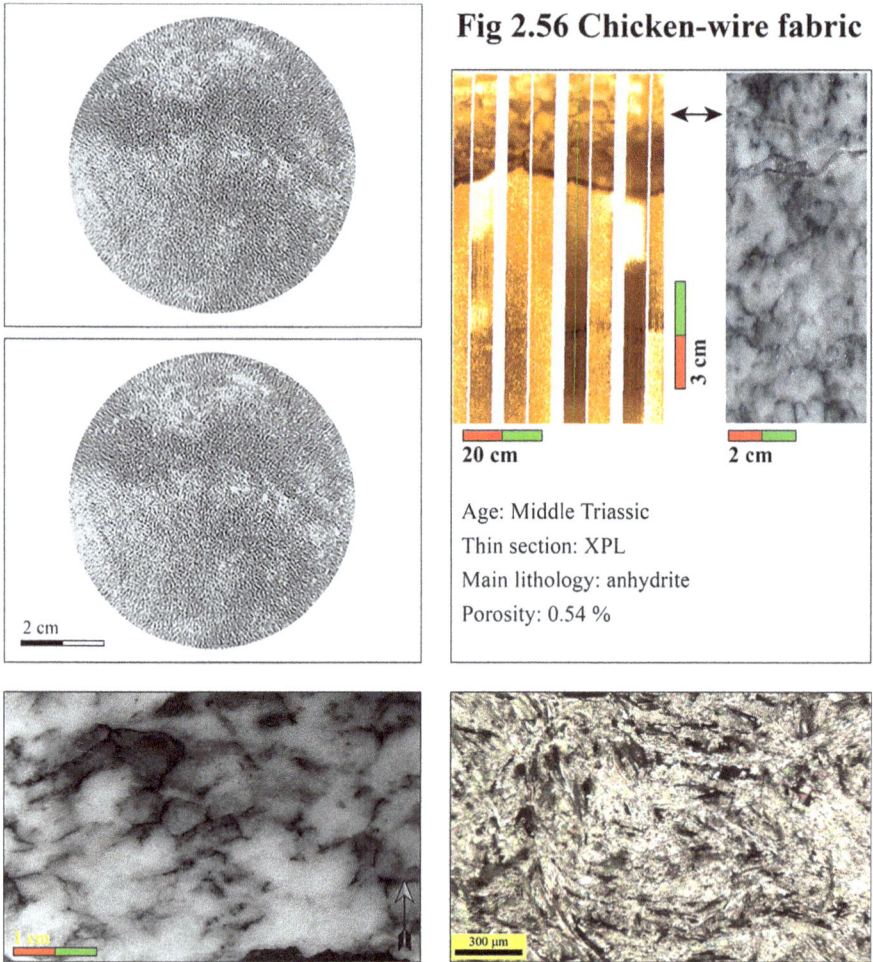

Fig 2.56 Chicken-wire fabric

20 cm

2 cm

3 cm

Age: Middle Triassic
Thin section: XPL
Main lithology: anhydrite
Porosity: 0.54 %

FIGURE 2.56
Chicken-wire fabric of anhydrite is visible in carbonate-evaporite environments. In CT scan images, this fabric can be recognized by patchy distribution of white jagged clasts with high frequency. They are also visible in FMI images (the upper part of the FMI log). Inclusions of dolomite and organic materials are present in many cases.

and aggregations, bioturbation and impregnation of sediments with organic materials (Figures 2.56–2.74). All of these CT photos are presented along with close-up and whole core photos as well as FMI and thin section images. Therefore, the reader can compare various proxies to understand the effect of these features on these images. Porosity value has been presented in all cases. Pore spaces are dark (low density) and so the porosity value can change the gray intensity of the CT image.

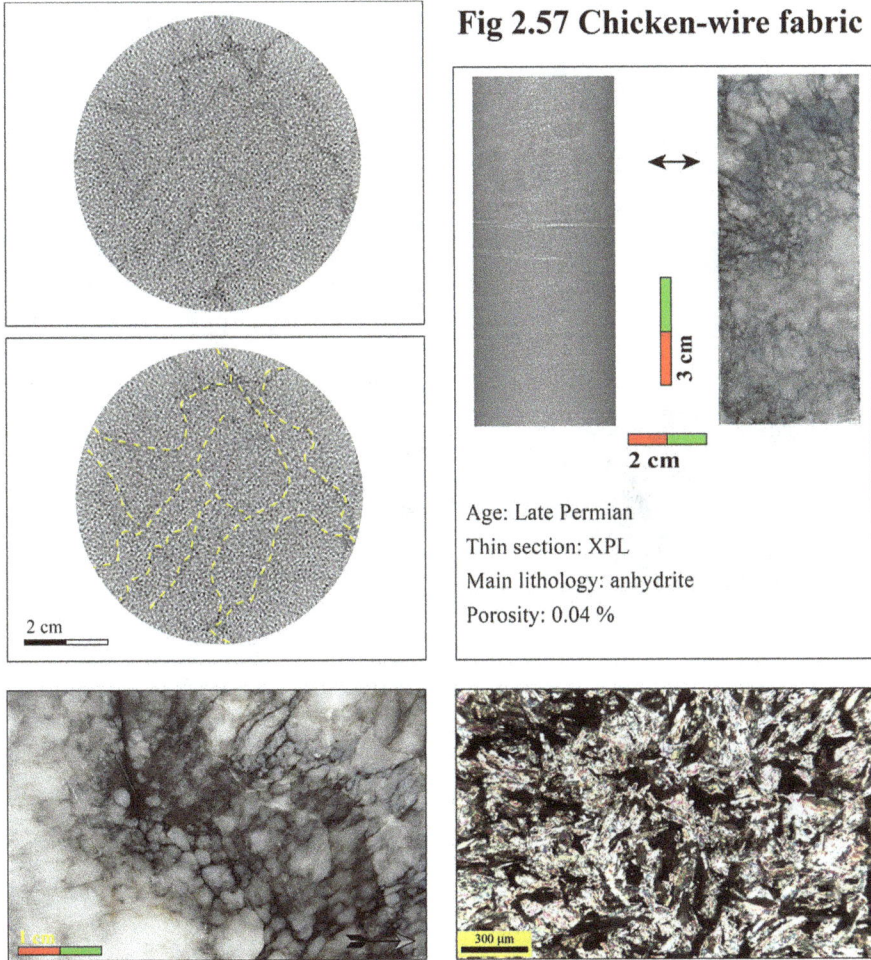

Fig 2.57 Chicken-wire fabric

Age: Late Permian

Thin section: XPL

Main lithology: anhydrite

Porosity: 0.04 %

FIGURE 2.57

Chicken-wire is a routine texture in anhydrites. It is composed of frequent anhydrite nodules with veins of carbonates (dolomites in most cases). Therefore, anhydrite nodules are seen as clasts in CT images (marked by yellow dashed lines). These clasts are accompanied with darker veins of dolomites. Usually, high anhydrite contents indicate deposition in shallow warm waters.

Fig 2.58 Clotted fabric

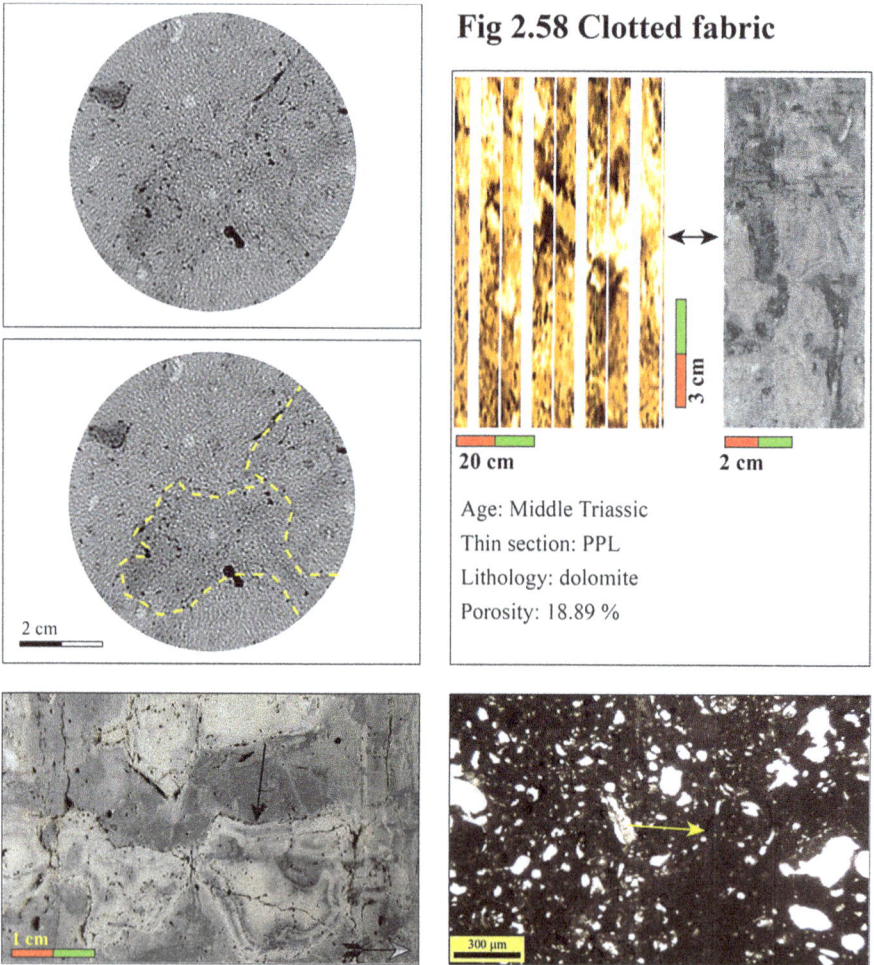

Age: Middle Triassic
Thin section: PPL
Lithology: dolomite
Porosity: 18.89 %

FIGURE 2.58

Clotted fabric is formed by organic materials in the carbonate sediments. The algae use the clasts as substrate and form a layered clotted fabric. This layering is more obvious at the outer parts. They can be recognized by their irregular shapes and lower densities (and so darker color) in CT images (dashed line). Low density is due to the high organic content. Some allochems (bioclasts debris) have been dissolved in later diagenetic stages. Arrows show the boundary of the clasts.

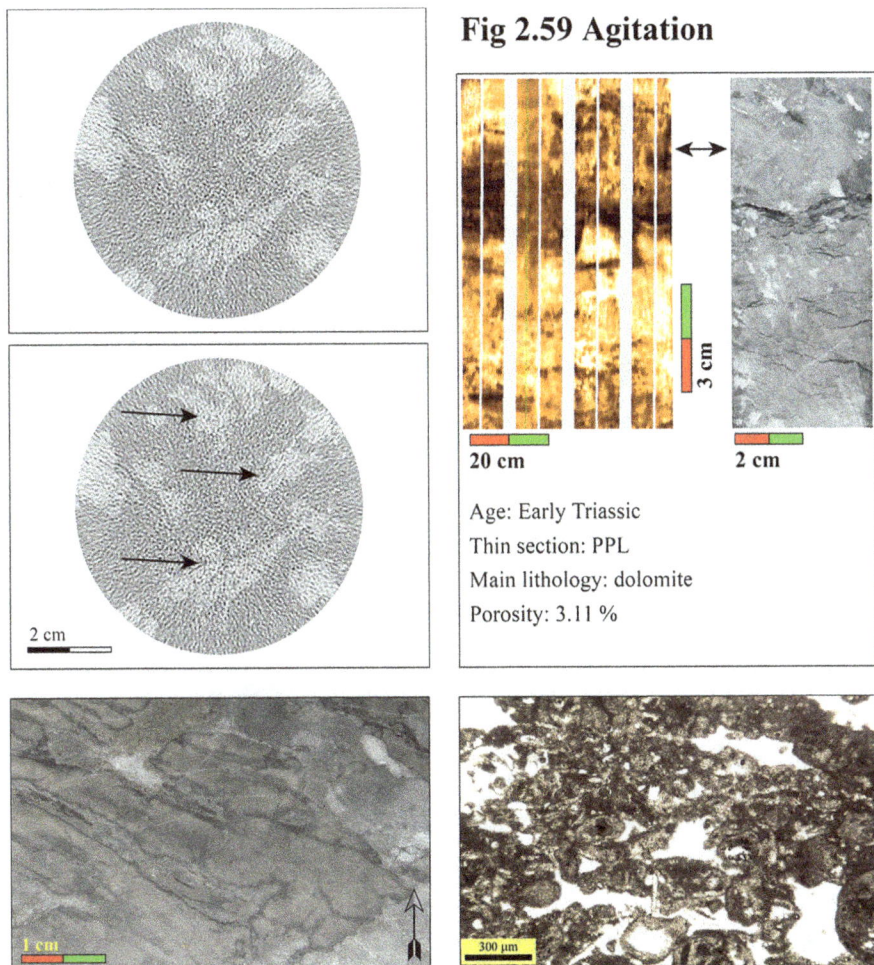

Fig 2.59 Agitation

Age: Early Triassic
Thin section: PPL
Main lithology: dolomite
Porosity: 3.11 %

FIGURE 2.59

Agitation is a routine process in shallow marine carbonates. Anhydrite is frequently precipitated and reworked again to form such deposits. They are easily recognized by large, bright, shapeless clasts in CT images (black arrows). A slump-like structure is visible in these images. High agitation is also evident in thin section and FMI images. High amounts of porosity are produced, which fill with anhydrite cement.

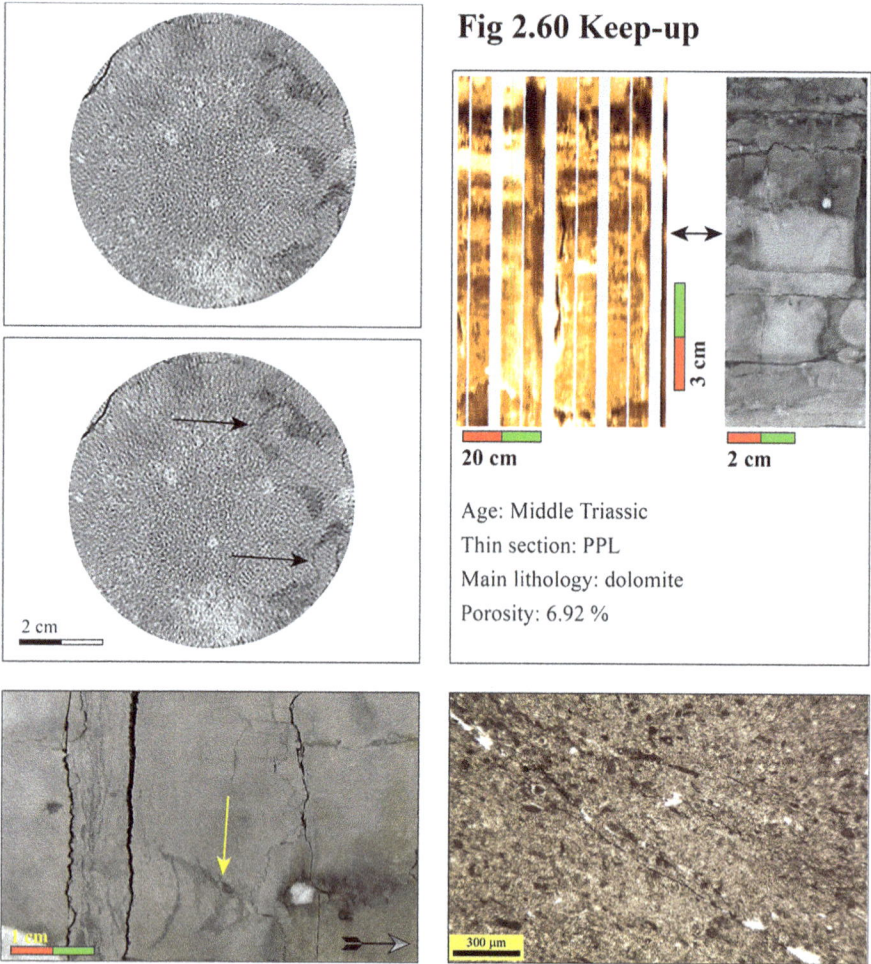

Fig 2.60 Keep-up

Age: Middle Triassic
Thin section: PPL
Main lithology: dolomite
Porosity: 6.92 %

FIGURE 2.60

Carbonates respond to the changes in the environmental conditions. The fauna keep up in response to sea-level rise to remain in approximately the same water depth. This process is seen in close-up core photo (yellow arrow) as well as CT images (black arrow). These are dome-like structures and so can be seen as curved surfaces in two-dimensional images. High organic content is also obvious in thin section studies.

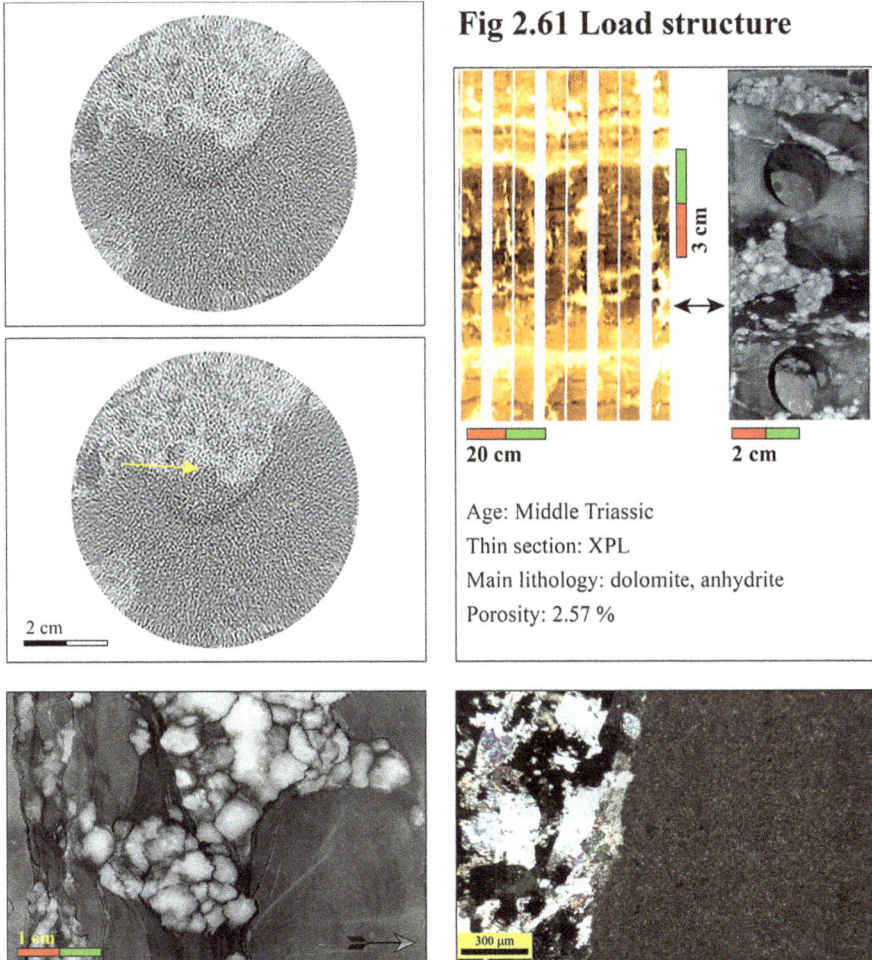

Fig 2.61 Load structure

Age: Middle Triassic

Thin section: XPL

Main lithology: dolomite, anhydrite

Porosity: 2.57 %

FIGURE 2.61

Anhydrite has higher density compared to dolomite or limestone at the time of deposition. Therefore, it moves downward into the lower sediments. The result is a load structure. A dome-like feature is formed. Therefore, it can be distinguished by a curved surface and lighter gray color in CT slices. This feature is also evident in core photo and FMI log. The dark lines in both close-up and CT images are stylolites.

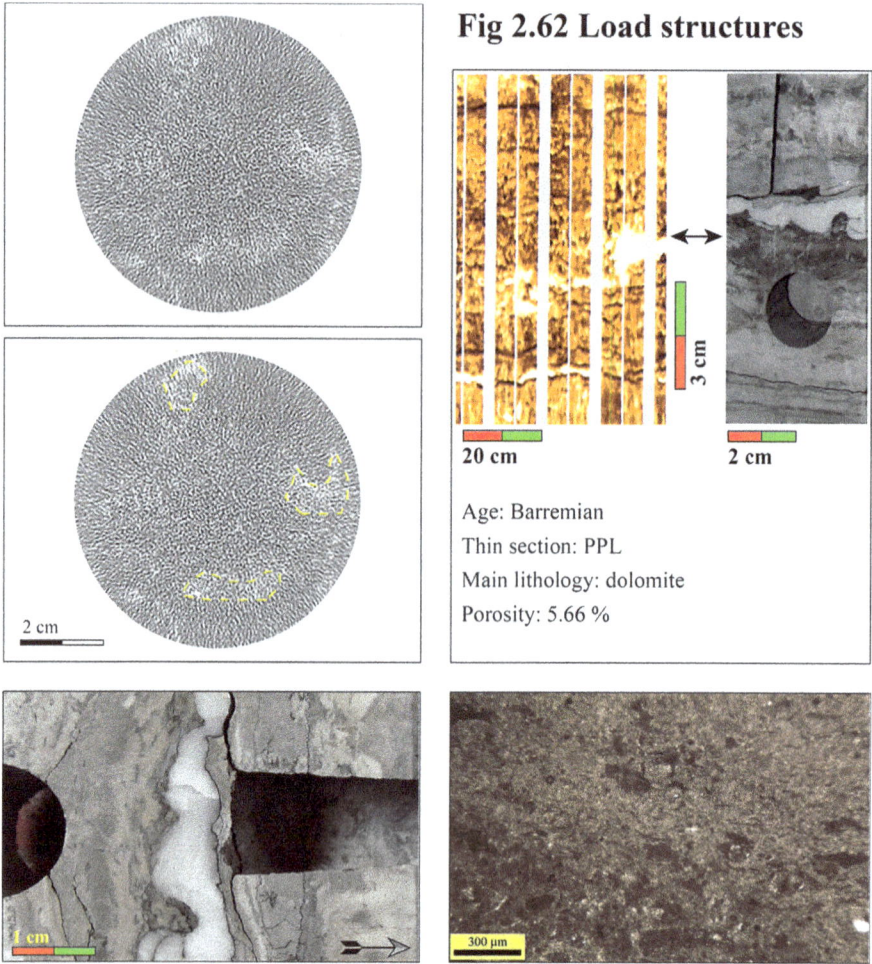

Fig 2.62 Load structures

20 cm   2 cm

Age: Barremian
Thin section: PPL
Main lithology: dolomite
Porosity: 5.66 %

2 cm

1 cm   300 µm

FIGURE 2.62

Load structures are the result of downward movement of dense minerals, especially anhydrite in carbonates. They can be observed as white closed surfaces in CT images (marked by yellow dashed lines). The sample has high anhydrite content and so light color is somehow scattered in the slice. Anhydrite mineral is marked by white color in FMI images.

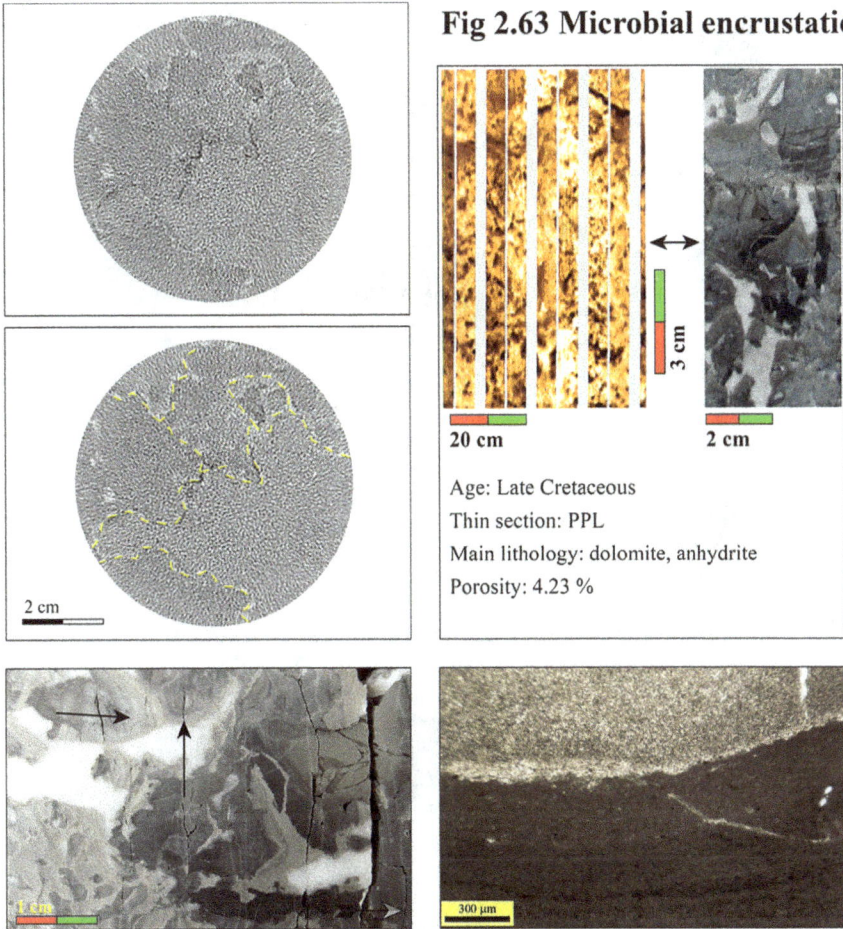

Fig 2.63 Microbial encrustation

Age: Late Cretaceous
Thin section: PPL
Main lithology: dolomite, anhydrite
Porosity: 4.23 %

FIGURE 2.63

Microbial organisms need a substrate to encrust and grow. They are routinely encrusted on grains. Due to high organic content and low density, these encrustations are recognized by dark colors in CT images. Large clasts have been encrusted by microorganisms in this sample. These thick layers can also be observed in core (arrows) and thin sections. Dark color is obvious in these photos.

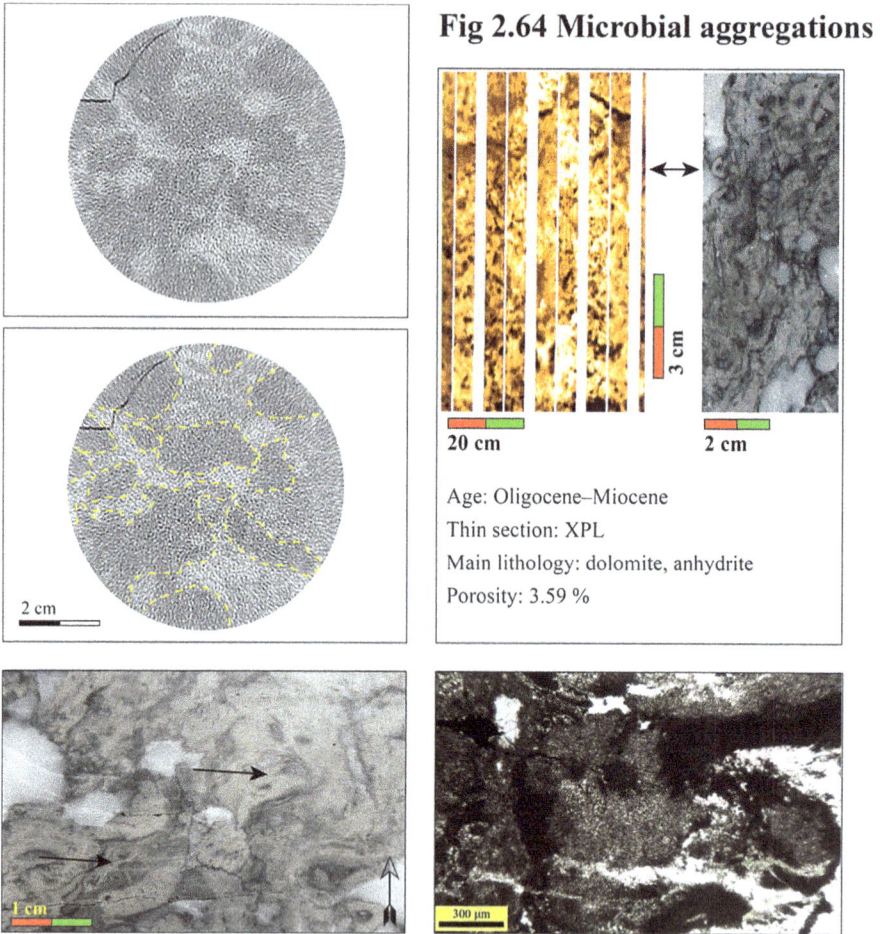

Fig 2.64 Microbial aggregations

Age: Oligocene–Miocene
Thin section: XPL
Main lithology: dolomite, anhydrite
Porosity: 3.59 %

FIGURE 2.64

Microbial aggregations are formed in shallow environments where enough food and oxygen are available. Here, they are recognized by their dark gray shades and various shapes (yellow dashed lines). These environments are also susceptible to evaporite deposition due to high rate of evaporation. Anhydrites are distinguishable with lighter color as compared to these structures.

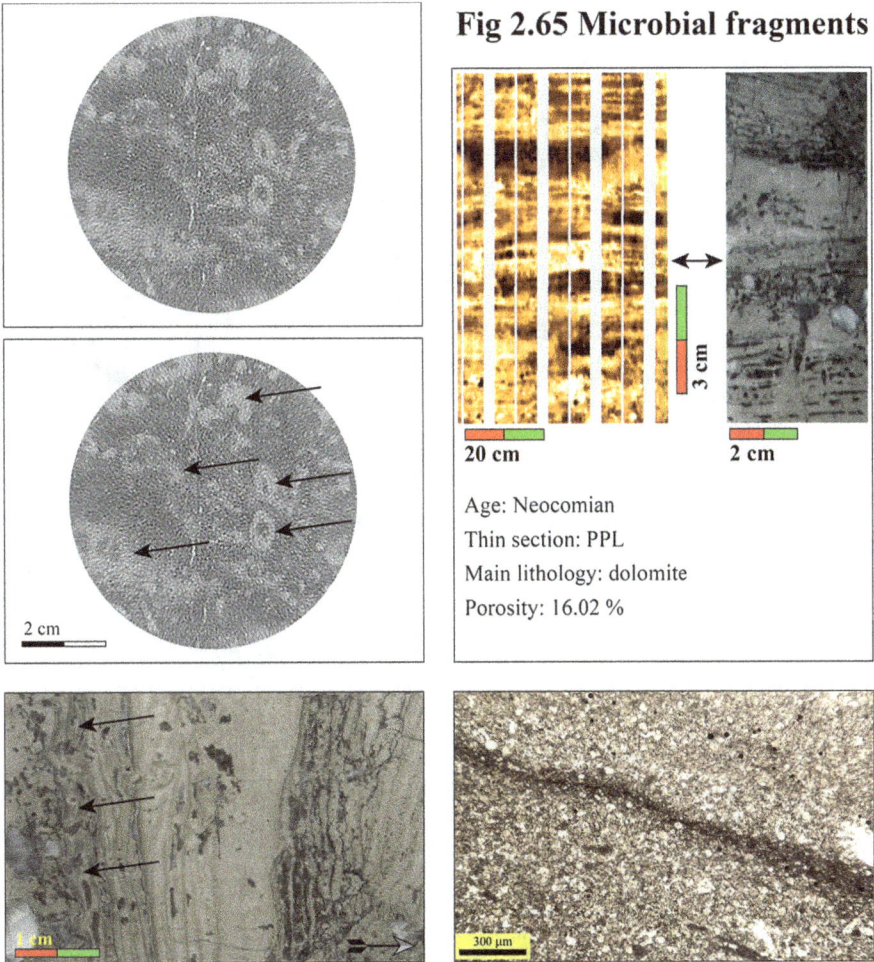

Fig 2.65 Microbial fragments

Age: Neocomian
Thin section: PPL
Main lithology: dolomite
Porosity: 16.02 %

FIGURE 2.65

Fragments of microbial colonies are obvious in this sample. With dolomite lithology (fabric-selective dolomitization), they can be marked by lighter color as compared to the limy background. As they are formed by living organisms, various shapes and sizes are visible (black arrows). They form fine laminations in sediments which are traceable in whole core, FMI and close-up photos.

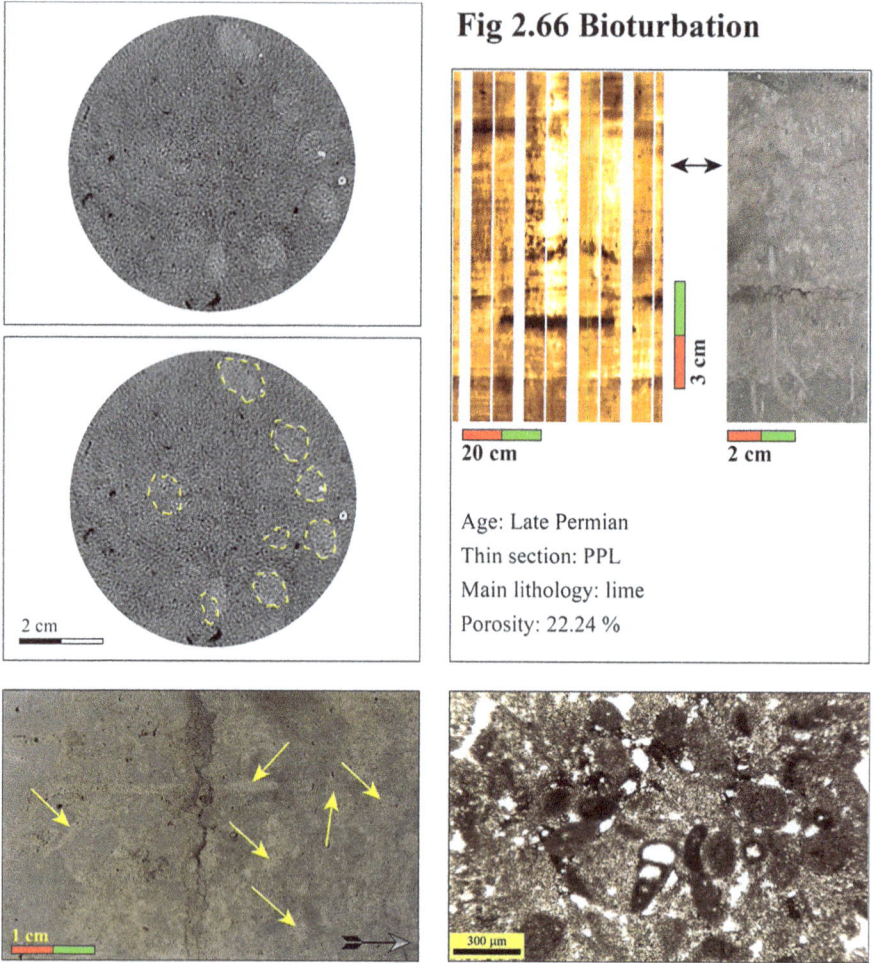

Fig 2.66 Bioturbation

Age: Late Permian
Thin section: PPL
Main lithology: lime
Porosity: 22.24 %

FIGURE 2.66

Bioturbations are traced in the rocks with different texture and/or remaining organic materials. Sometimes, porosity increases due to this process. As the lithology is the same, they are hardly traceable on CT images. Here, bioturbations have been formed in different directions and their cross sections are visible on core and CT slices with lighter color (higher anhydrite content).

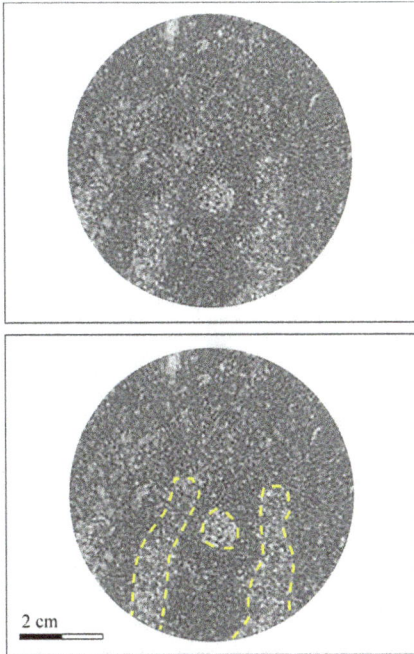

Fig 2.67 Boring

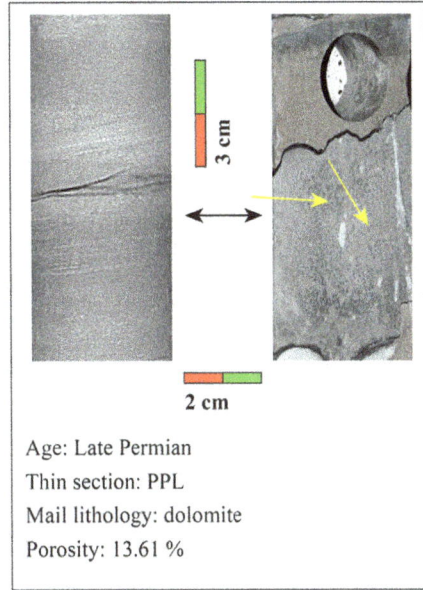

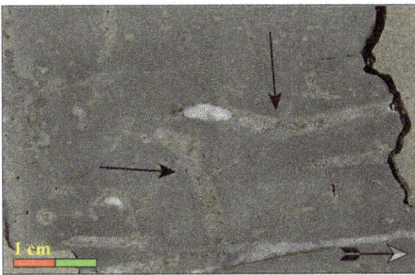

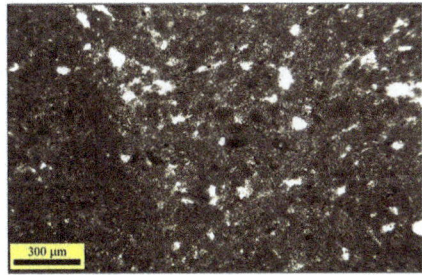

Age: Late Permian

Thin section: PPL

Mail lithology: dolomite

Porosity: 13.61 %

FIGURE 2.67

Boring is the result of fauna activities in sedimentary rocks. They can be seen as tube-like structures or circles in CT images (dashed lines), based on their orientation. In most cases, they are filled with higher density materials (close up and whole core photos, arrows) and so are evident with lighter color in CT images. Sometimes, the porosity is higher in these structures and they are darker than the matrix, but the shapes of cross-sections are always the same.

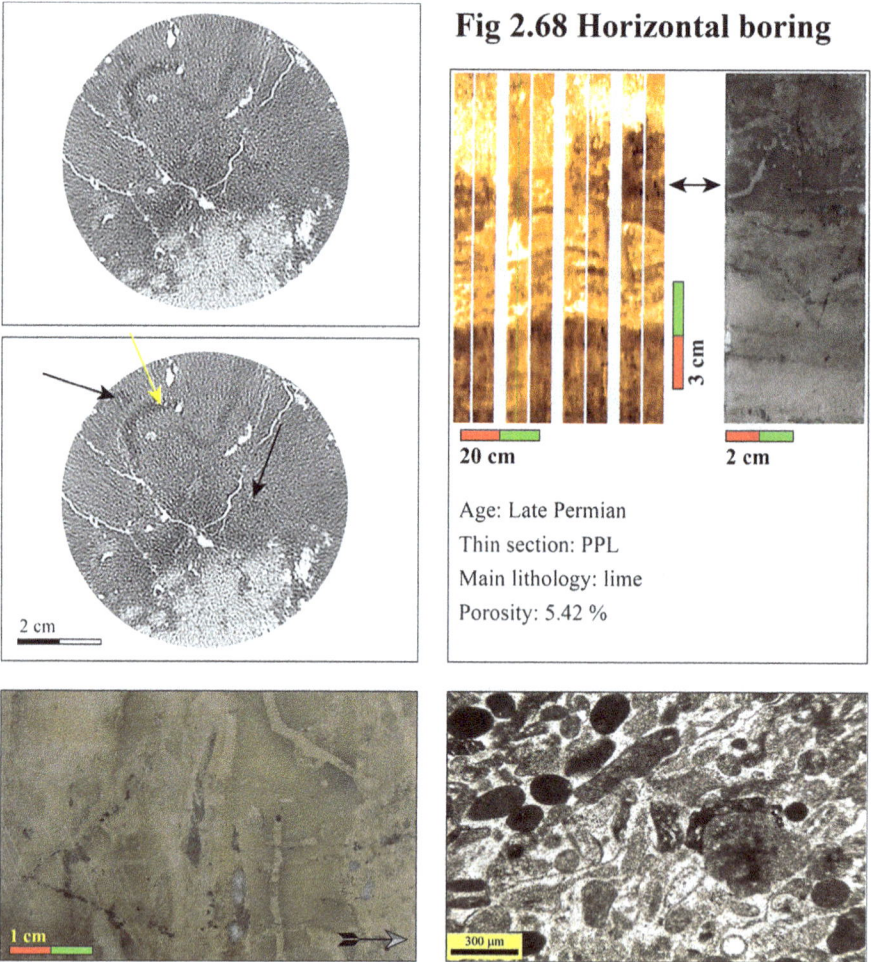

Fig 2.68 Horizontal boring

Age: Late Permian

Thin section: PPL

Main lithology: lime

Porosity: 5.42 %

FIGURE 2.68

Borings are rarely seen in CT images. Here, an s-shaped horizontal boring (yellow arrow) is visible in the CT slice. There are also some other borings but their transversal sections (black arrows) are seen as circles and therefore cannot be recognized easily. These borings fill with different materials compared to matrix in many cases. Therefore, their traces in CT images are visible.

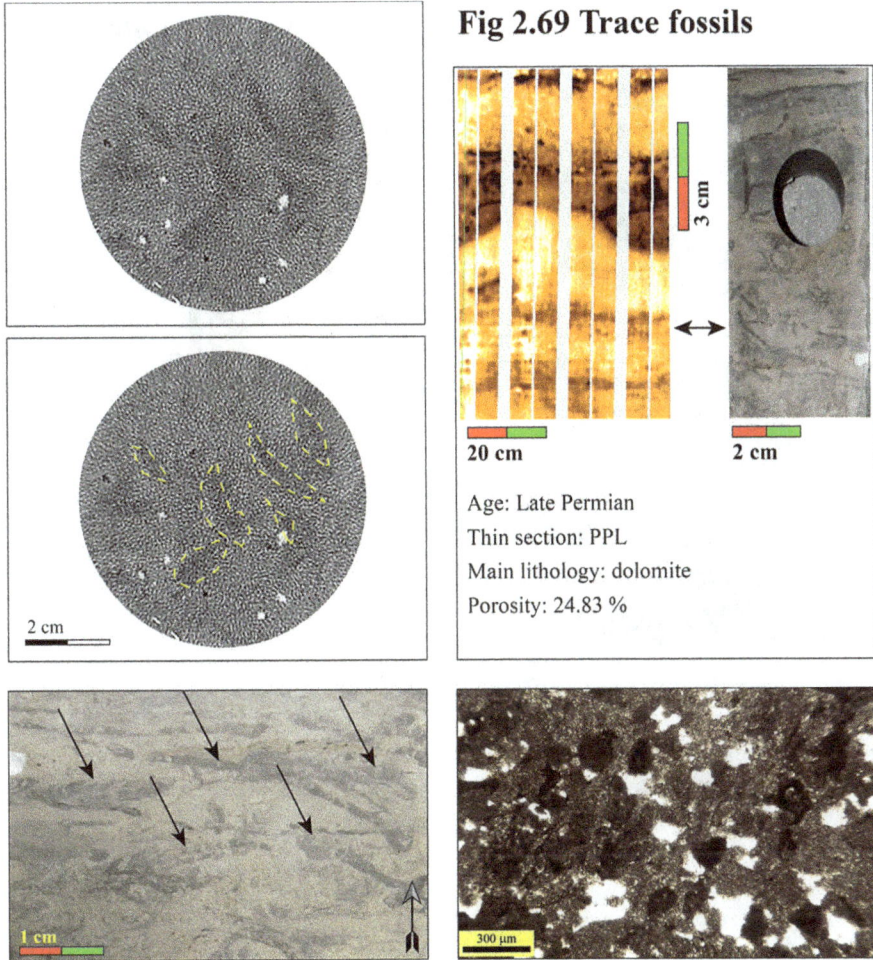

Fig 2.69 Trace fossils

Age: Late Permian
Thin section: PPL
Main lithology: dolomite
Porosity: 24.83 %

FIGURE 2.69

Sediments have been re-distributed by the living organisms. These traces are recognized by different colors in CT images. They are usually darker than background because contain more porosity. These structures are also visible on core (arrows). Their patchy and irregular forms help to distinguish them on CT slices.

Fig 2.70 Layer boundary

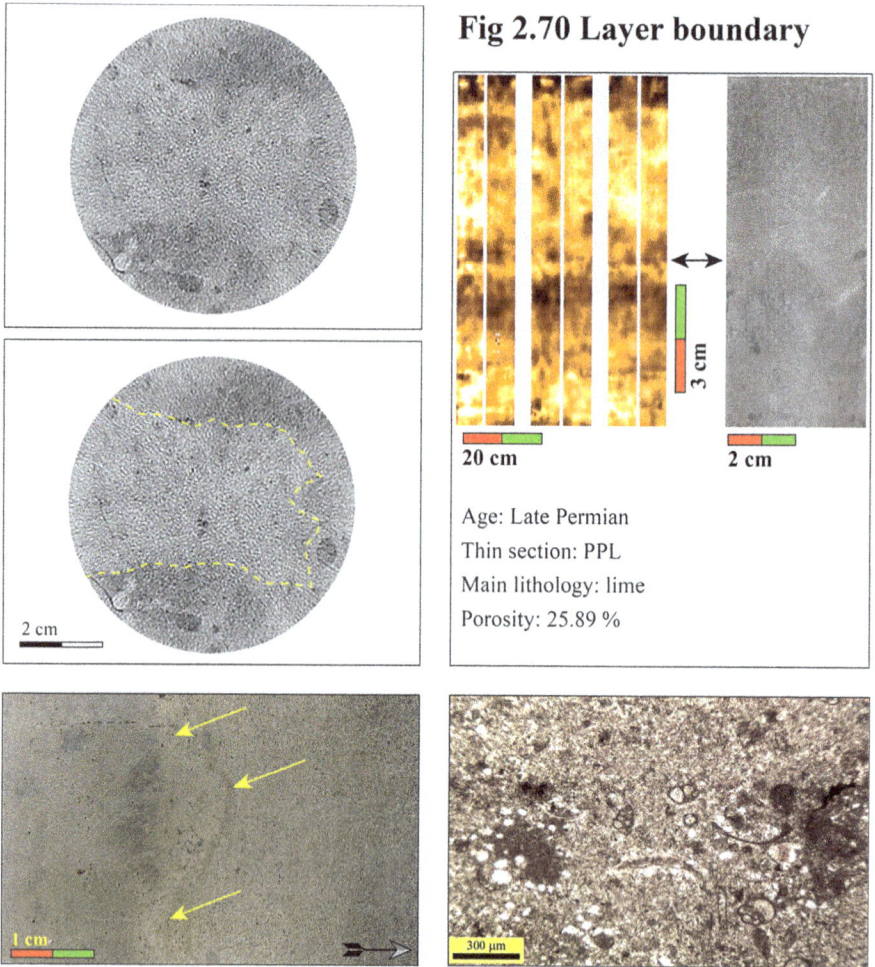

20 cm   2 cm

Age: Late Permian
Thin section: PPL
Main lithology: lime
Porosity: 25.89 %

FIGURE 2.70
A dome-like boundary between two layers can be observed as a filled curve with lighter shade of gray in CT image. The layers are differentiated by their texture (grainstone and packstone) which is evident on close-up photo of the core (yellow arrows). Lithology is pure lime and so change in the texture of two layers creates such contrast. The grainstone layer has more porosity.

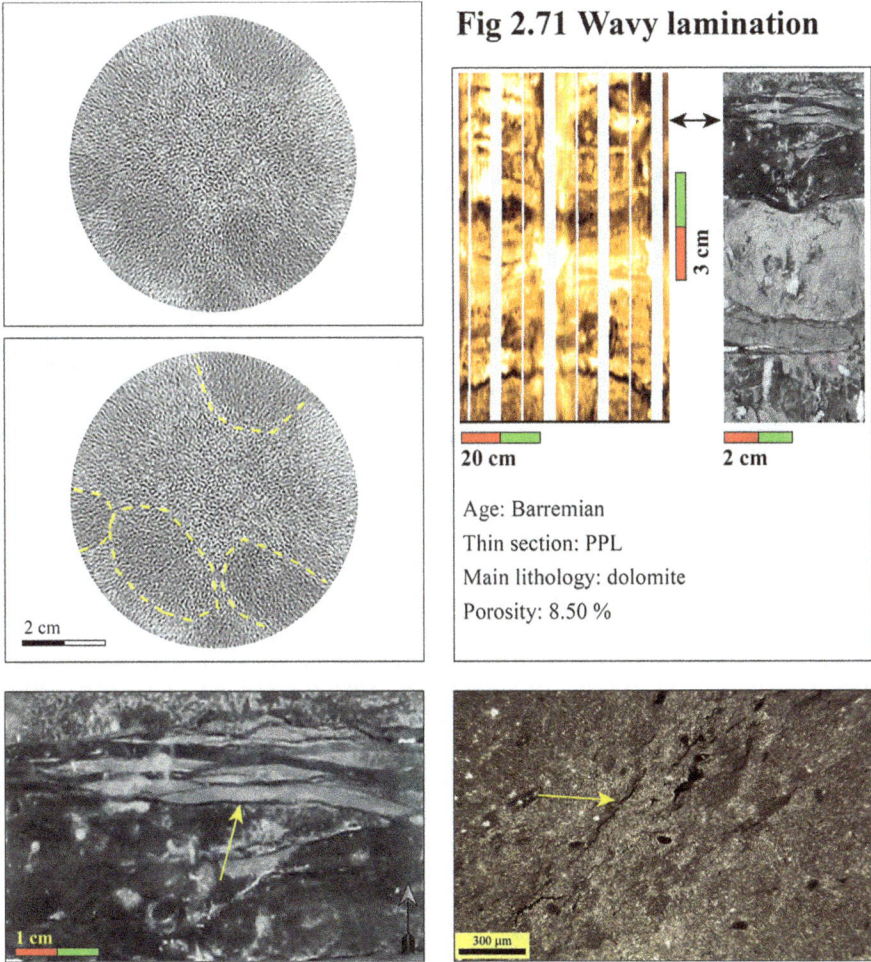

Fig 2.71 Wavy lamination

Age: Barremian
Thin section: PPL
Main lithology: dolomite
Porosity: 8.50 %

FIGURE 2.71

Wavy laminations are not frequent in carbonate depositional environments. Their dome-like shape can be distinguished by large closed areas in CT images (dashed lines). Depositional conditions were similar for all layers and therefore difference in gray colors is negligible. Anyway, some insoluble materials such as clays and iron oxides have been accumulated in the boundary of the layers (the arrows).

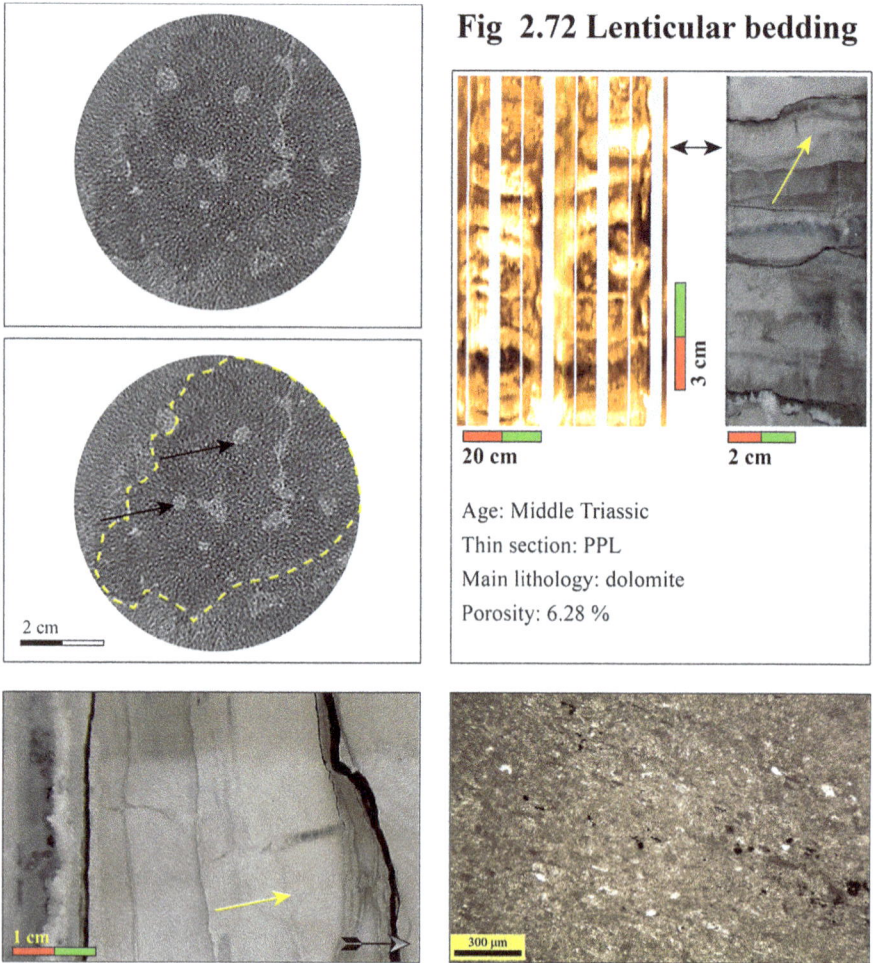

Fig 2.72 Lenticular bedding

20 cm   2 cm

Age: Middle Triassic
Thin section: PPL
Main lithology: dolomite
Porosity: 6.28 %

FIGURE 2.72
Because of their large sizes, lenticular beddings are rare at core scale. These pictures show a lenticular bed in mudstones (yellow arrow) of peritidal environment. This bed has lower density (higher porosity or organic content) and so is marked by dark color in the CT image. Rounded milky white circles are transversal cross sections of borings with high anhydrite content (black arrows). Layering is evident in core, thin section and FMI photos, but it is not observable on CT image.

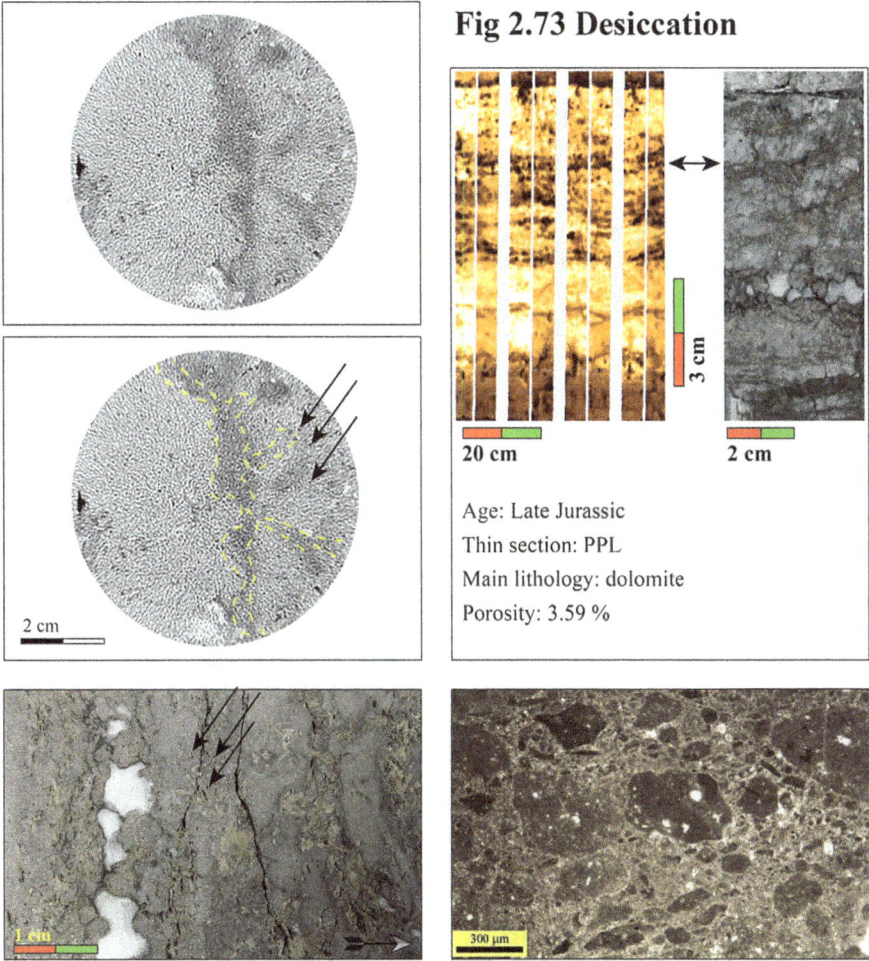

Fig 2.73 Desiccation

Age: Late Jurassic
Thin section: PPL
Main lithology: dolomite
Porosity: 3.59 %

FIGURE 2.73
Desiccation and de-watering structures are formed in arid environments. They usually indicate a subaerial exposure. These surfaces contain high organic content and porosity and therefore show darker color in CT images. Here, the main surface and cracks are visible (yellow dashed lines). Pores are marked by arrows. Generally, the CT image is light in color, due to high anhydrite content.

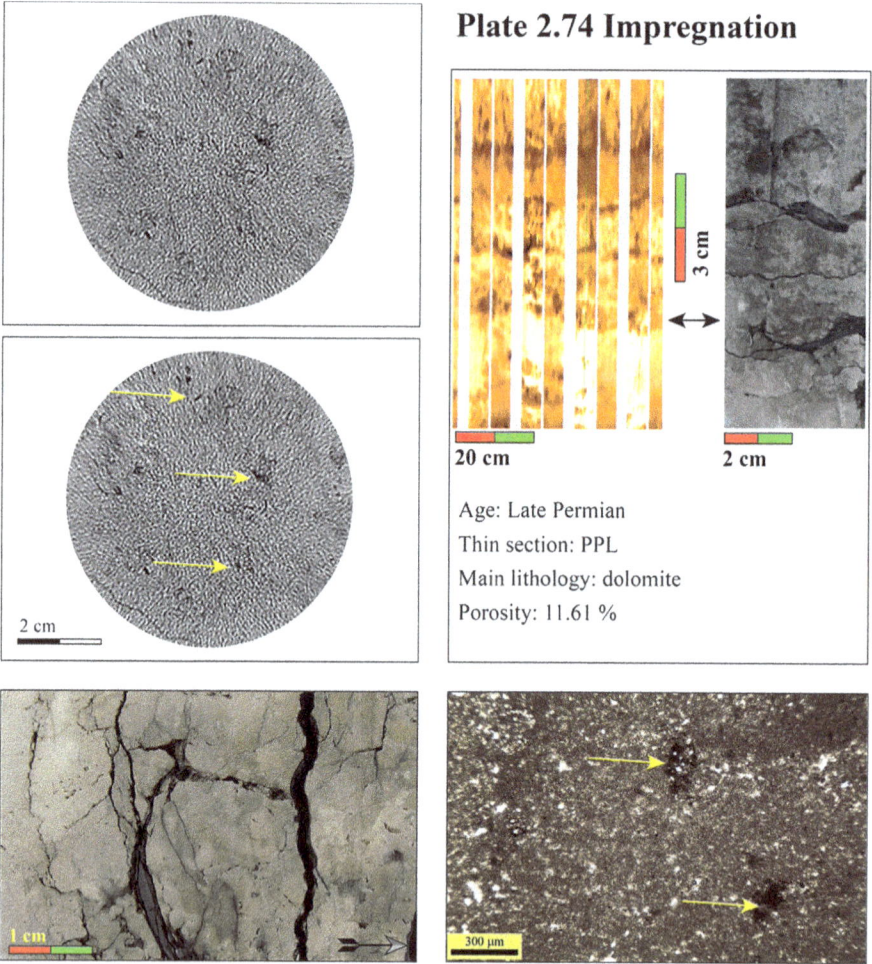

Plate 2.74 Impregnation

Age: Late Permian
Thin section: PPL
Main lithology: dolomite
Porosity: 11.61 %

FIGURE 2.74

Change in the gray intensity is obvious but no special grain or texture could be recognized. This is due to the natural impregnation of sediments with clays, organic matters and opaque minerals. Such impregnation is also evident on thin section image (yellow arrow) and close-up photo. Desiccation cracks are visible on close-up and CT image (lower left). Main lithology is dolomite with minor amounts of anhydrite.

3

Planar Structures

3.1 Introduction

Planar structures are two-dimensional surfaces which can be seen in three-dimensional space of the reservoir. They are formed by depositional processes (e.g., bedding and cross bedding), later deformation of the rocks (e.g., faults and fractures) or combination of both (e.g., solution seams and stylolites). These structures include faults and fractures, scour surfaces, stylolites, solution seams, laminations, bedding and cross-bedding surfaces. They change the heterogeneity and reservoir quality of the carbonate reservoirs. Planar structures are seen as lines with various properties in CT scan images. The color and shape of the line strongly depend on the type of surface as well as its filling materials.

Among all these structures, fractures are more important because they have strong effects on the reservoir quality and production of the hydrocarbon-bearing formations. A significant part of the world reservoirs is fractured or at least fractures change the reservoir properties of these formations. Recent advances in CT scan image acquisition significantly increased our understanding of these reservoirs. Fracture modeling using various data types (e.g., CT scan images, FMI log, core description and production data) can explain the reservoir production and predict future reservoir behavior. This chapter presents the effect of planar structures in CT images. These images are valuable sources for fracture characterization in subsurface studies. CT images are prepared before core opening and therefore fractures have not been affected by this process. Using many examples, various types of fractures are described and characterized in this chapter.

The most common planar structure in sedimentary rocks is layering (Figures 3.1–3.4). This feature indicates any change in sediment properties, which include physical, chemical and biological conditions of the depositional environment. The first four figures of this chapter show the effect of layer boundaries on CT images. As rock properties are different in different layers or laminas, they are distinguishable in these images. It also depends on the angle between the surface and CT slice. Horizontal surfaces are not

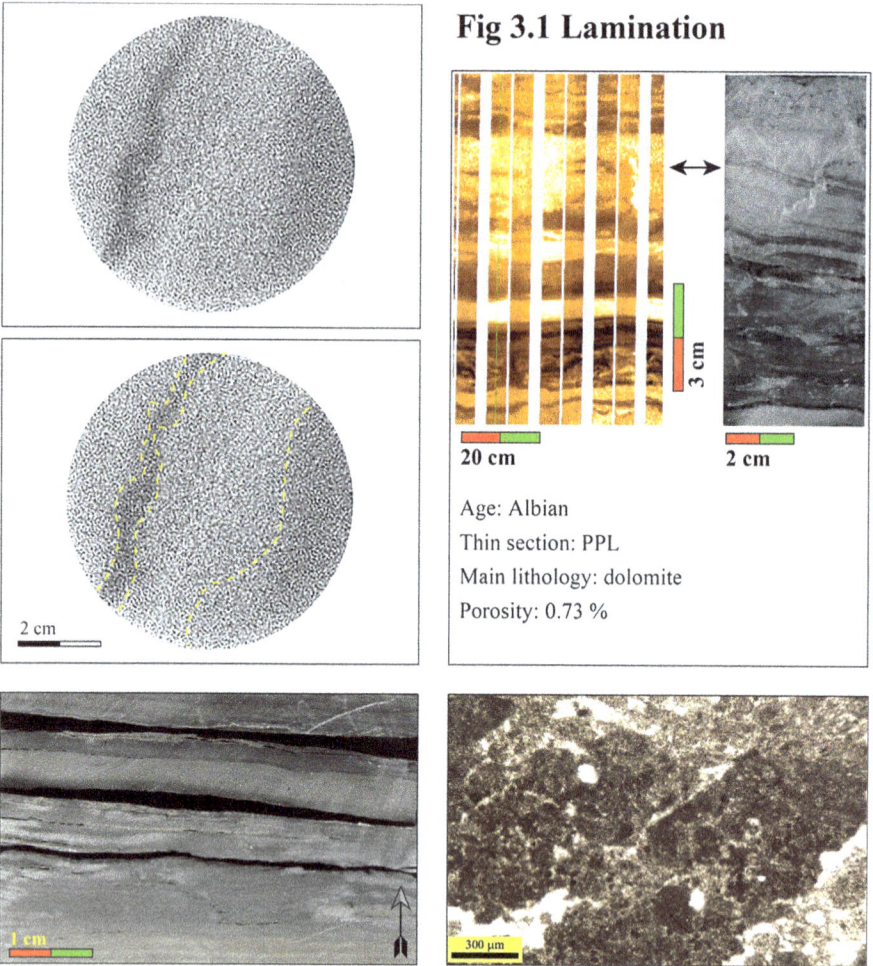

Fig 3.1 Lamination

20 cm   2 cm

Age: Albian

Thin section: PPL

Main lithology: dolomite

Porosity: 0.73 %

FIGURE 3.1

Horizontal laminations are not detectable in CT images of vertical wells, as the slices are horizontal too, in most cases. Sometimes, layers are not horizontal. Therefore, they are marked by alternation of dark and light gray colors (dashed lines) in CT slices. This is due to the change in material density in various layers.

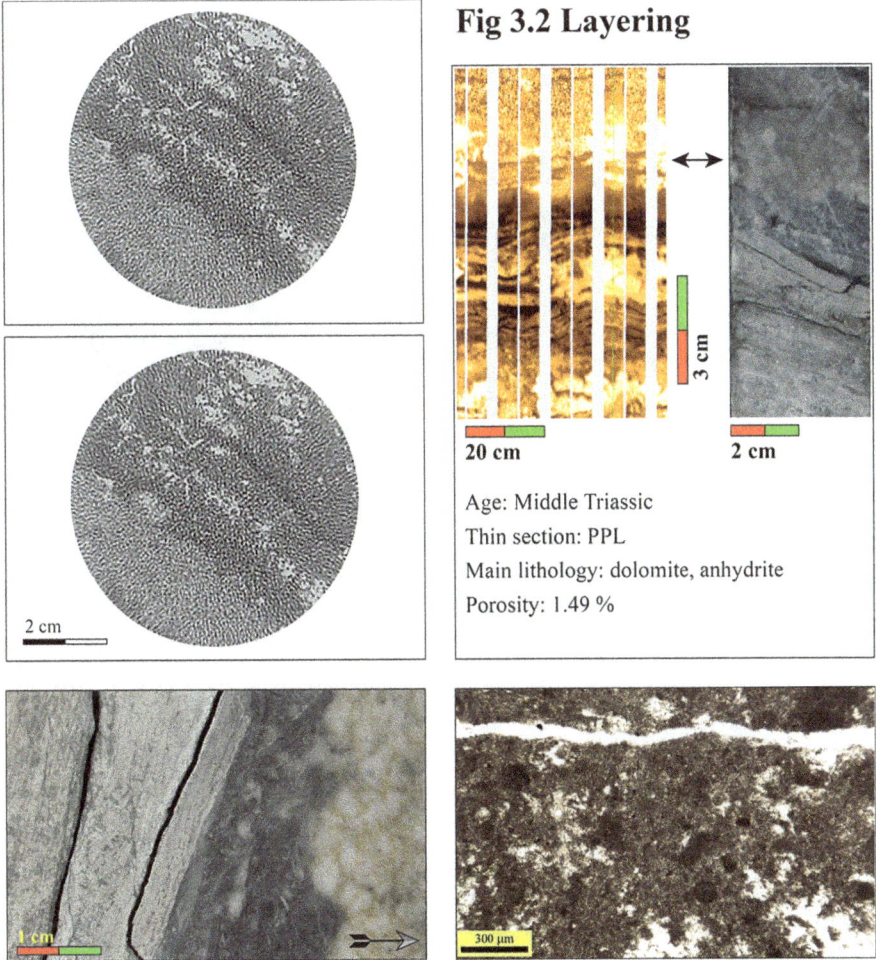

Fig 3.2 Layering

Age: Middle Triassic
Thin section: PPL
Main lithology: dolomite, anhydrite
Porosity: 1.49 %

FIGURE 3.2

Layering is a very common sedimentary structure in all kinds of sediments. Layers are differentiated by different sediment characteristics. Densities of the materials are also different in many cases. Therefore, they are marked by different bands of gray intensity in CT images. This is also the case about this sample. Different anhydrite content cause layering in this sample.

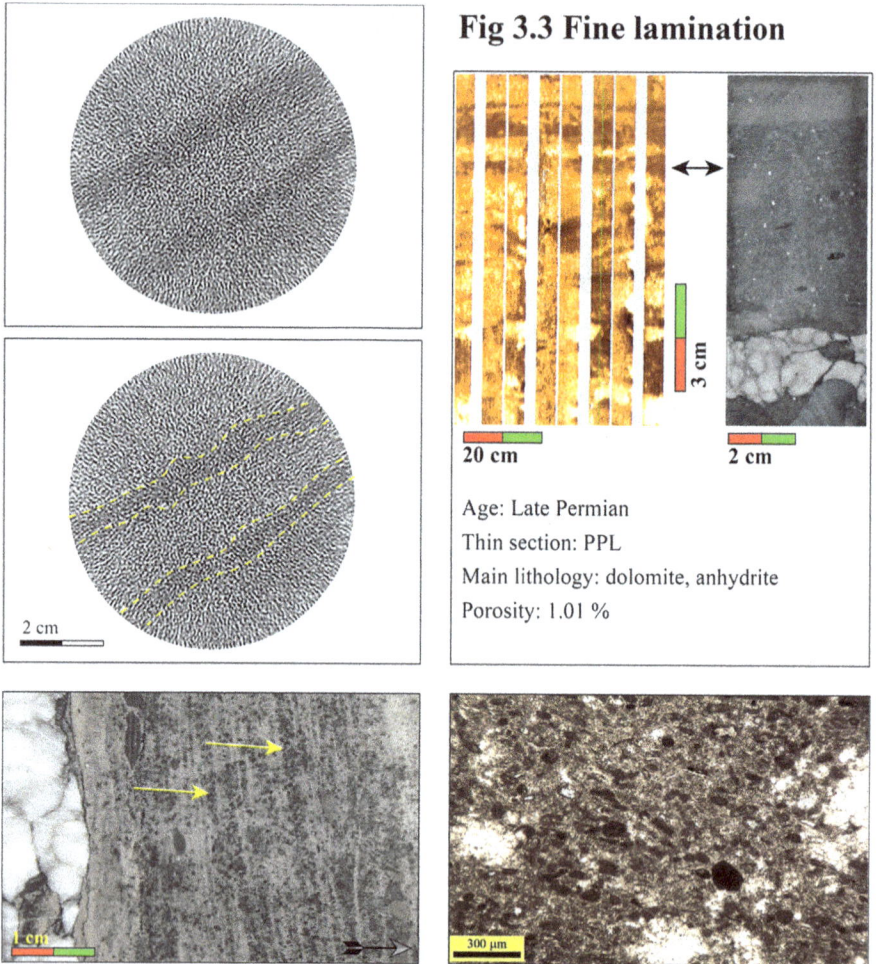

Fig 3.3 Fine lamination

20 cm   **2 cm**

Age: Late Permian

Thin section: PPL

Main lithology: dolomite, anhydrite

Porosity: 1.01 %

FIGURE 3.3

Laminations indicate changes in depositional conditions. They are marked as lines with different thickness in CT slices (dashed lines). The widths of the lines depend on the thickness of the layers. Pellets have been accumulated in some layers in this sample (arrows) and therefore they are traced by darker gray intensity in CT images. These pellets are composed of micrites which contain high microporosity.

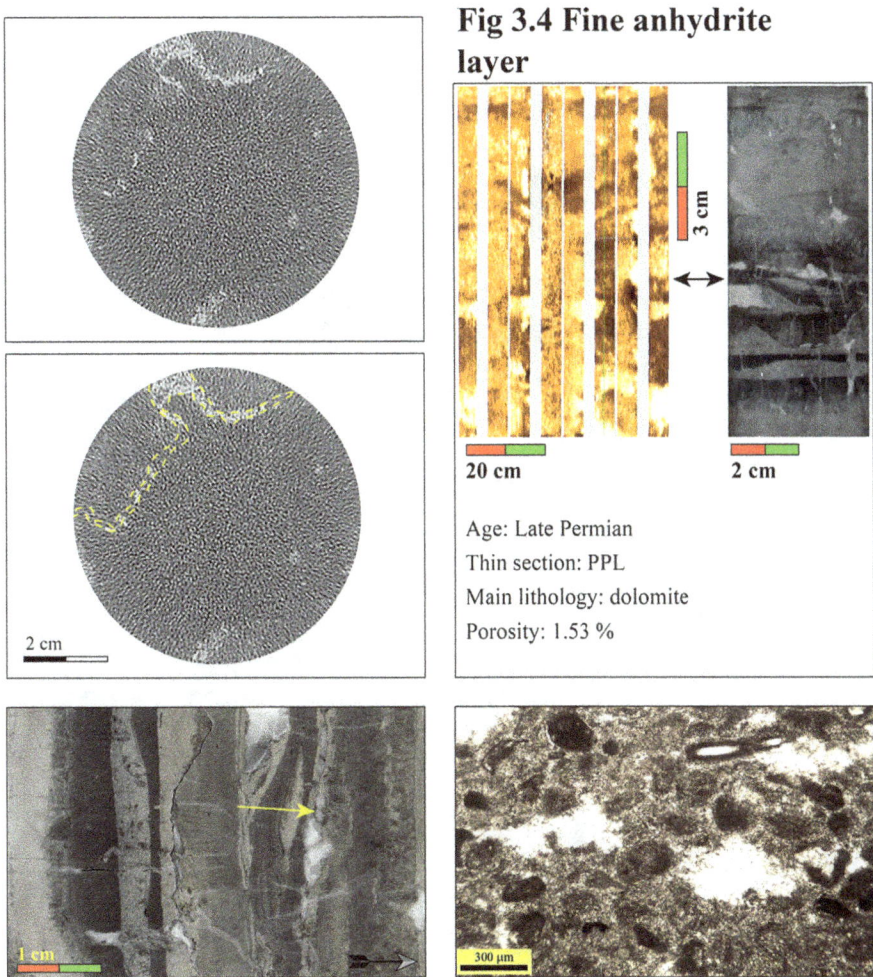

Fig 3.4 Fine anhydrite layer

Age: Late Permian
Thin section: PPL
Main lithology: dolomite
Porosity: 1.53 %

FIGURE 3.4
Generally, layers are marked as lines with darker or lighter gray color in CT images. As anhydrite is denser than other carbonate minerals (calcite and dolomite), it is distinguishable with light gray (milky white) in CT slices. Here, anhydrites are present as fine layers (dashed line on CT and arrow on close-up photos). The thickness of the layer varies at the core scale.

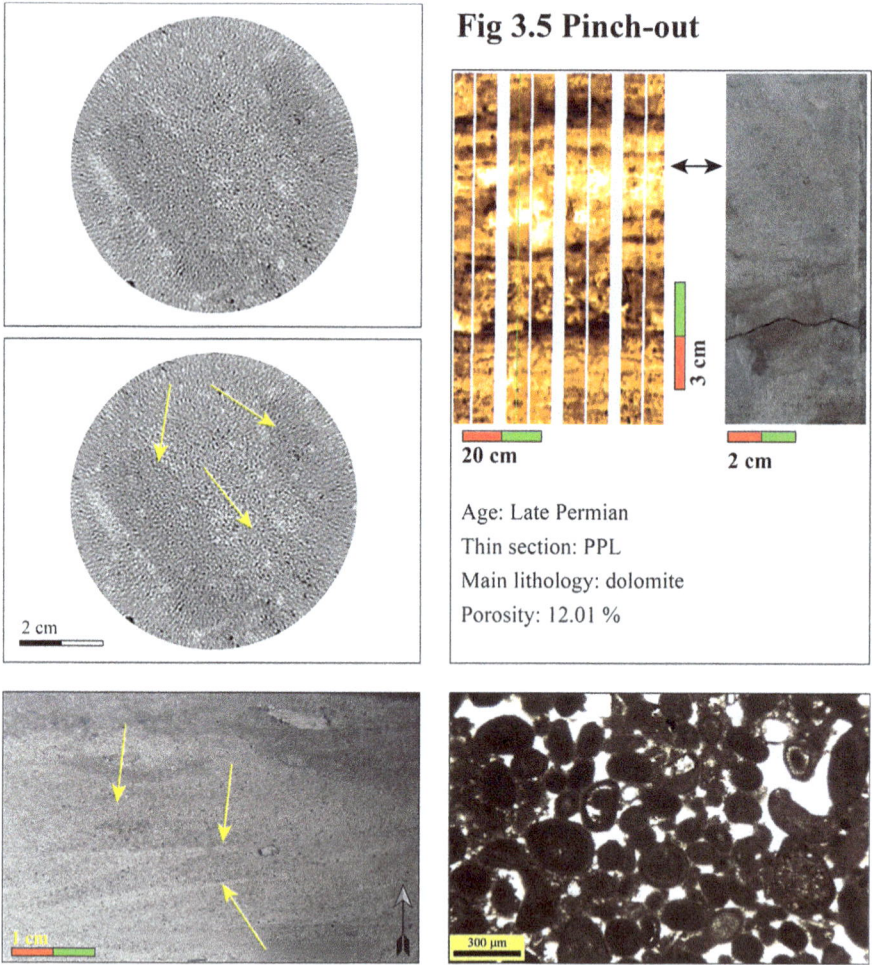

Fig 3.5 Pinch-out

Age: Late Permian

Thin section: PPL

Main lithology: dolomite

Porosity: 12.01 %

FIGURE 3.5

Layers are pinched-out in transition zones of sedimentary environments. These structures are not visible at core scale in many cases. Here, small-scale pinched-outs are visible in close-up and CT images (arrows). Layers have different properties and so their densities are different. Some layers contain more porosities in these cases.

detectable in horizontal CT slices. Therefore, a minimum angle from horizon is required to detect them in these images. Sometimes, these layers are pinched-out at the core scale. This is very uncommon but is illustrated in Figure 3.5 of this book. Also, the boundaries of the layers are erosional in some cases. This erosion shows reactivation of the environmental energy and can be used in sequence stratigraphy and reservoir zonation. Therefore, it is

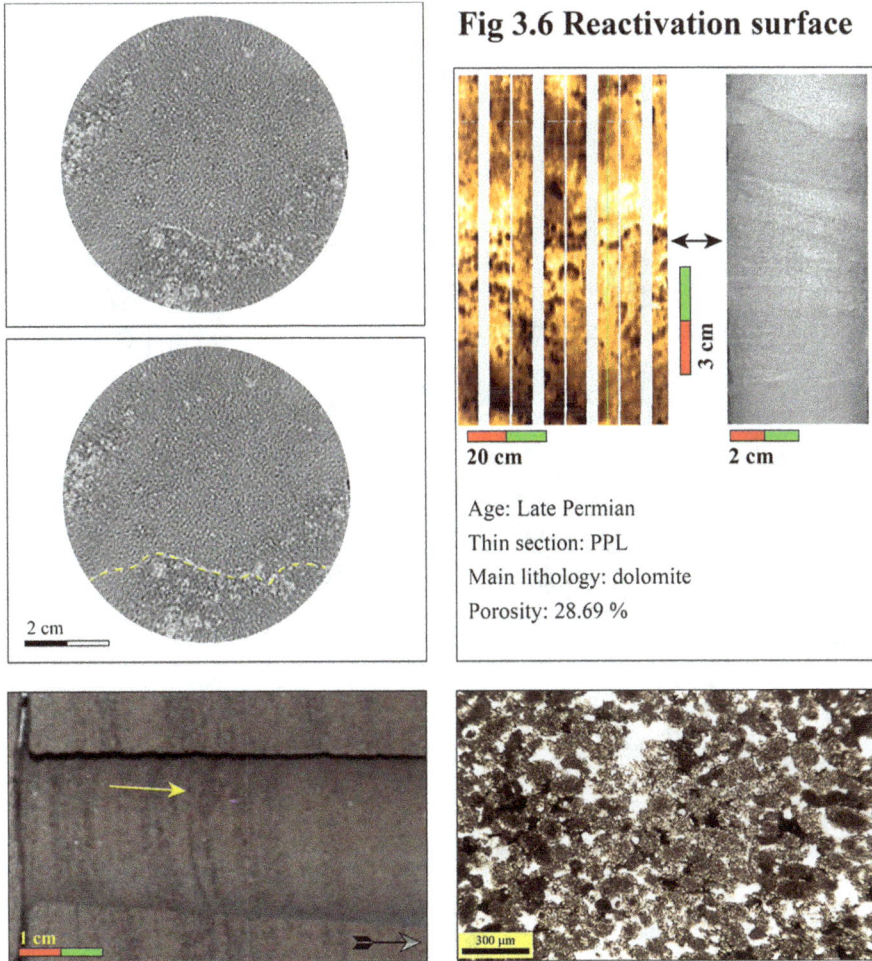

Fig 3.6 Reactivation surface

Age: Late Permian
Thin section: PPL
Main lithology: dolomite
Porosity: 28.69 %

FIGURE 3.6
Reactivation surfaces indicate change in the energy of the depositional environment. An erosional surface is formed which is observable on transverse (dashed line) and longitudinal CT image as well as FMI and core (arrow). Here, a reactivation surface has been created in the cross-bedded ooid grainstone facies. Dolomitized peloids and intraclasts can be recognized with light gray color in the CT images.

important to trace them in CT photos. Figures 3.6–3.7 of this chapter present such surfaces and their trace in CT images.

As mentioned earlier, layers are not horizontal in many cases. This may be due to the primary depositional conditions or secondary tectonic activities. Cross-beddings and cross-laminations are formed in depositional environments. Layers are deposited at an angle to the main bedding plane.

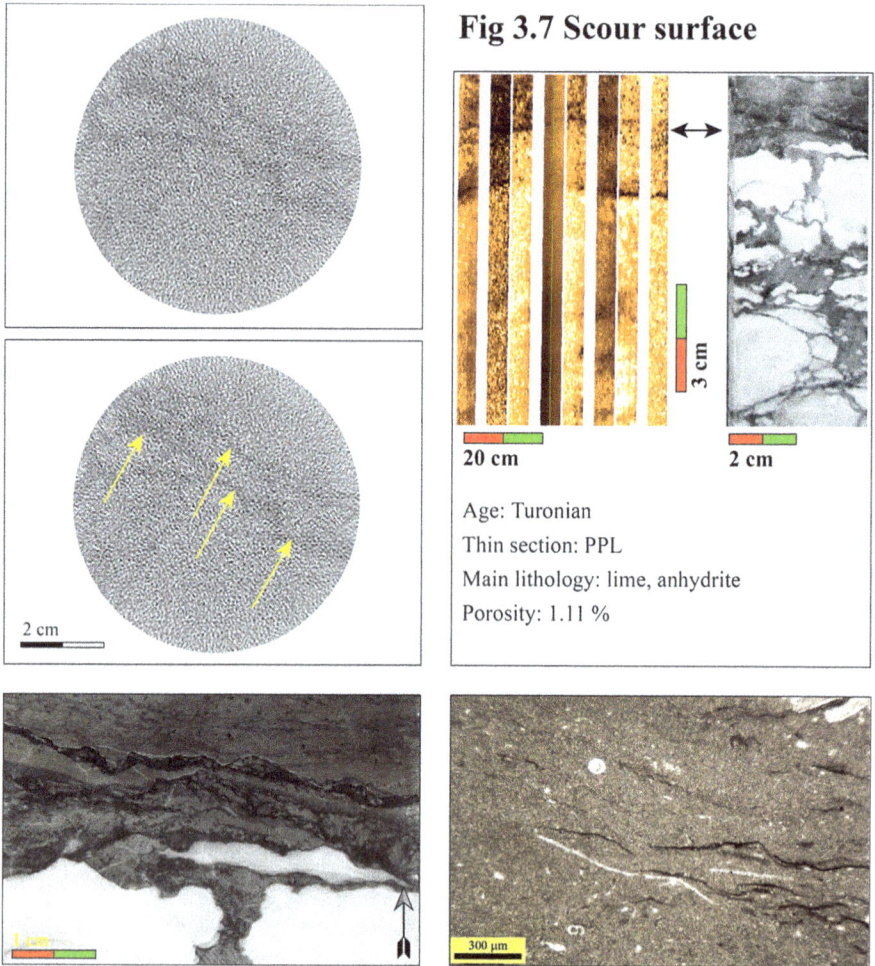

Fig 3.7 Scour surface

20 cm   2 cm

Age: Turonian
Thin section: PPL
Main lithology: lime, anhydrite
Porosity: 1.11 %

FIGURE 3.7

Water erodes the sediments when environmental energy increases. In this case, a scour surface is created. The lower facies are enriched in clays because they have been deposited at the lowest energy level. Also, insoluble materials such as clays and organic materials are enriched in these surfaces as they resist against further dissolution. Therefore, these surfaces are traced as dark, irregular lines in CT images (arrows), like stylolites, but they are generally lighter in color.

The result is a set of parallel and inclined layers. This structure shows the effect of the fluid flow. As layers are not horizontal, they can be seen in CT scan images. Such as other types of layers, sediments' characteristics are different in successive layers. Examples include mineralogy, porosity, cement type and frequency. Therefore, they have different densities and are marked

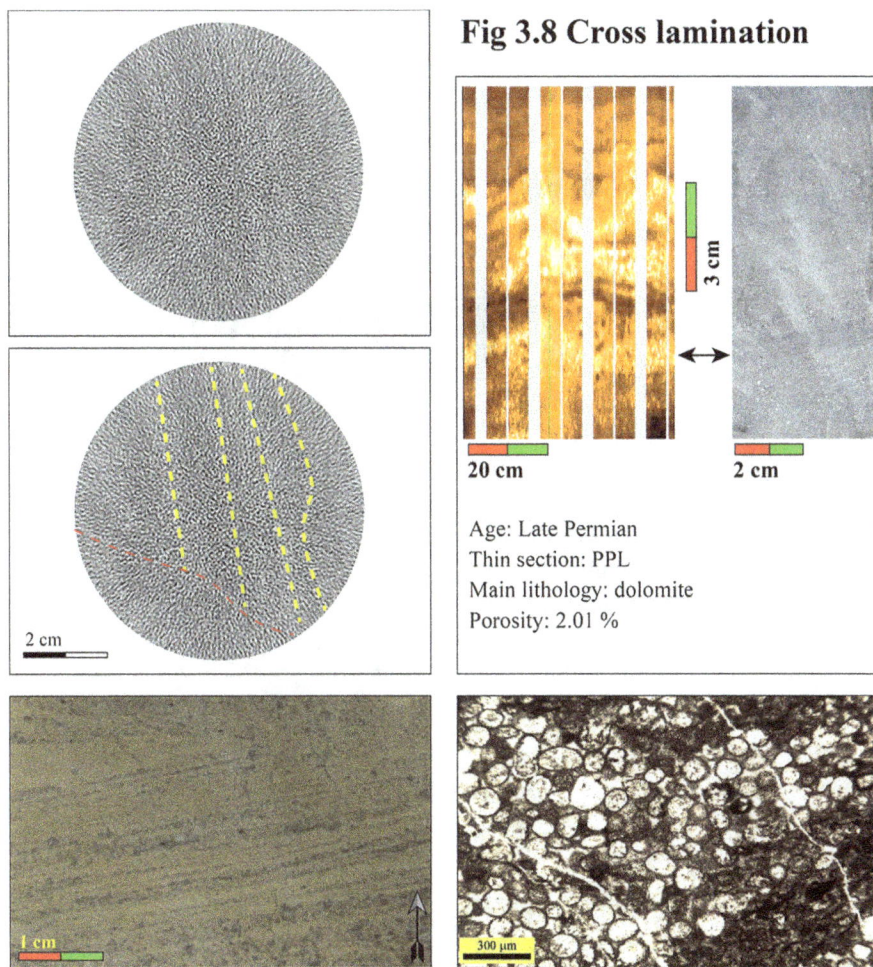

Fig 3.8 Cross lamination

Age: Late Permian
Thin section: PPL
Main lithology: dolomite
Porosity: 2.01 %

FIGURE 3.8
Cross laminations are visible in CT image (yellow dashed lines) of a dolomitic interval. Laminations have been created by anhydrite filled ooids and micritized small peloids. A hairline filled fracture is also observable in CT image (red dashed line) which truncates the laminations. These laminations are also evident in FMI (wavy lines), whole core, close-up and thin section images of the sample.

as striped lines in CT images. They are not straight in some cases. Figures 3.8–3.15 of this chapter show various types of cross-beddings with different effects in CT images.

Sediments are compacted after deposition. At first, physical compaction affects the sediments and decreases porosity and permeability. Then, the effects of chemical compaction can be seen as solution seams and stylolites.

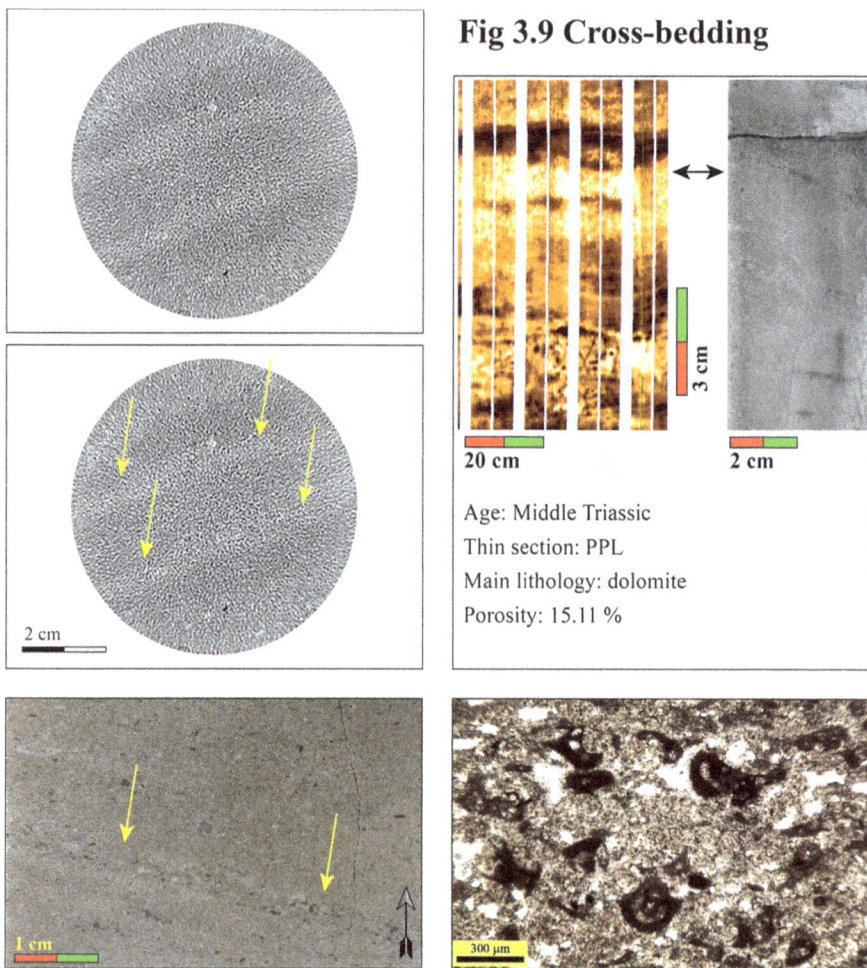

Fig 3.9 Cross-bedding

Age: Middle Triassic
Thin section: PPL
Main lithology: dolomite
Porosity: 15.11 %

FIGURE 3.9

Some layers are not parallel with the planes of general stratification of the formation, which are usually horizontal. They show the flow direction in many cases. Cross-beddings are routinely formed in terrigenous rocks, but sometimes allochems are oriented and form these structures in carbonates. They are marked by alternation of light and dark gray colors in CT slices (arrows). They are also visible in whole core, close-up and FMI images.

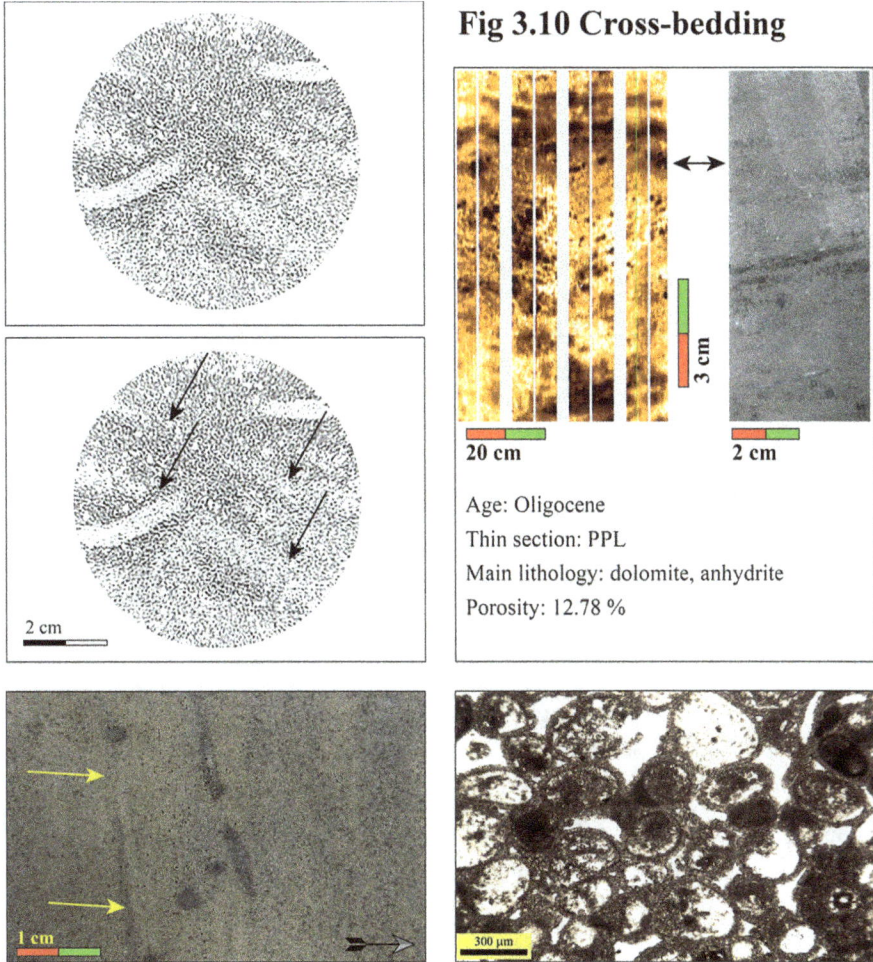

Fig 3.10 Cross-bedding

Age: Oligocene
Thin section: PPL
Main lithology: dolomite, anhydrite
Porosity: 12.78 %

FIGURE 3.10

Cross-beddings are formed in high-energy, grain-dominated environments in carbonate formations. They can be traced as alternation of light and dark bands in CT images (yellow arrows).
The intensity of gray color depends on the difference in density of adjacent layers. Here, high-
porosity layers are seen as dark gray bands of the image, while low-porosity layers are lighter
in color (black arrows).

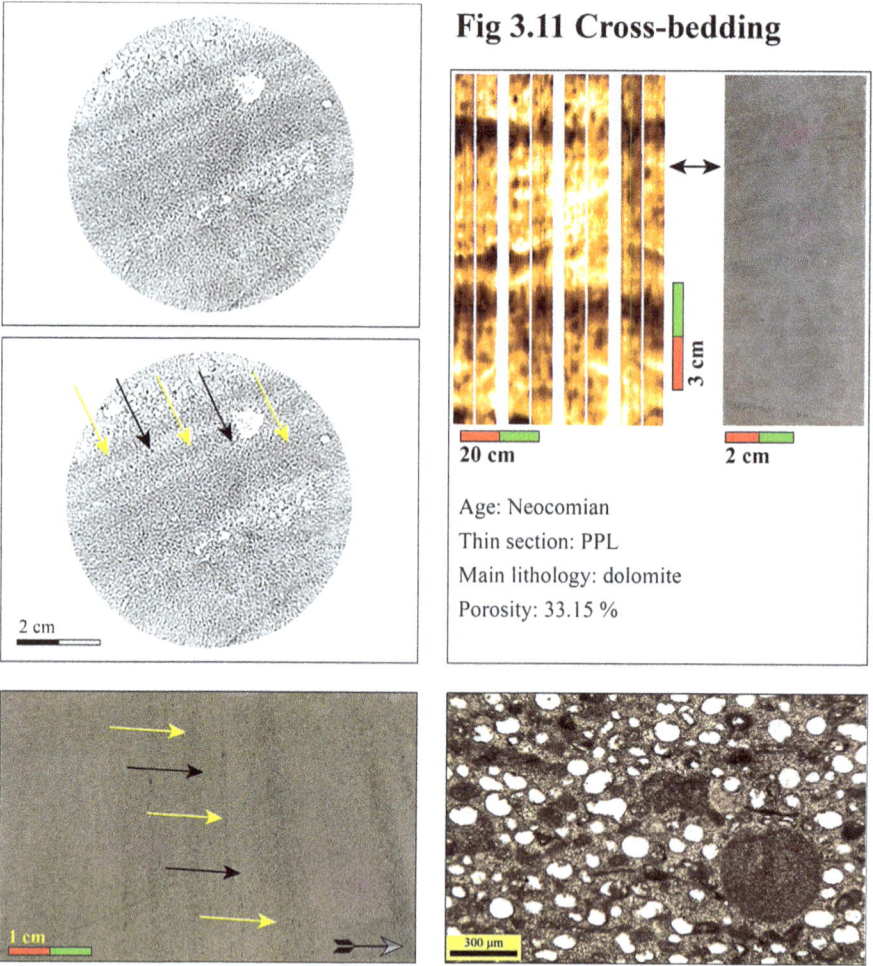

Fig 3.11 Cross-bedding

Age: Neocomian

Thin section: PPL

Main lithology: dolomite

Porosity: 33.15 %

FIGURE 3.11

A clear alternation of dark- and light-color cross-bedded layers (arrows) is obvious in these pictures. It seems that both porosity and mineralogy are different in these layers. Pores are moldic and can be seen as dark circles in CT image. Thick layers are also visible in FMI image. Some pores have been filled with anhydrite cement.

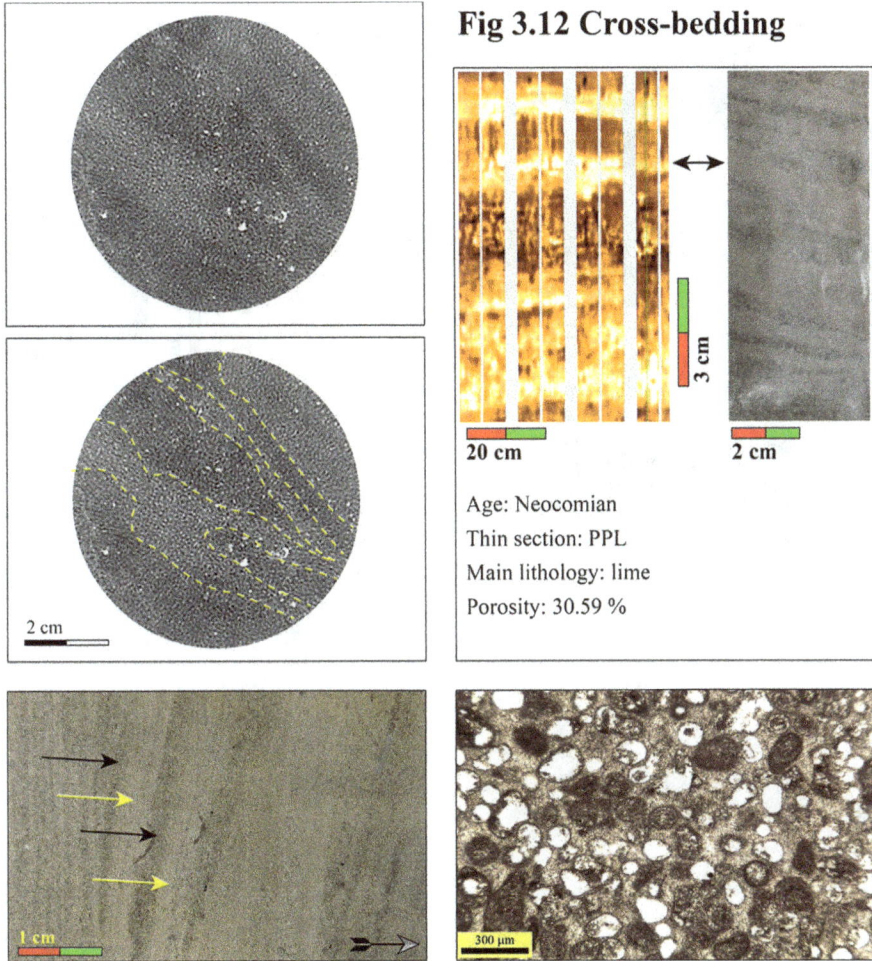

Fig 3.12 Cross-bedding

Age: Neocomian
Thin section: PPL
Main lithology: lime
Porosity: 30.59 %

FIGURE 3.12

Layers with different colors are the result of variations in depositional conditions. Such changes are visible in CT slices (dashed lines), whole core, close-up (black and yellow arrows) and FMI images of this cross-bedded carbonate core. Moldic pores are present in the entire sample but their frequency changes in various layers. In fact, these pores are molds of ooids, which are responsible for layering.

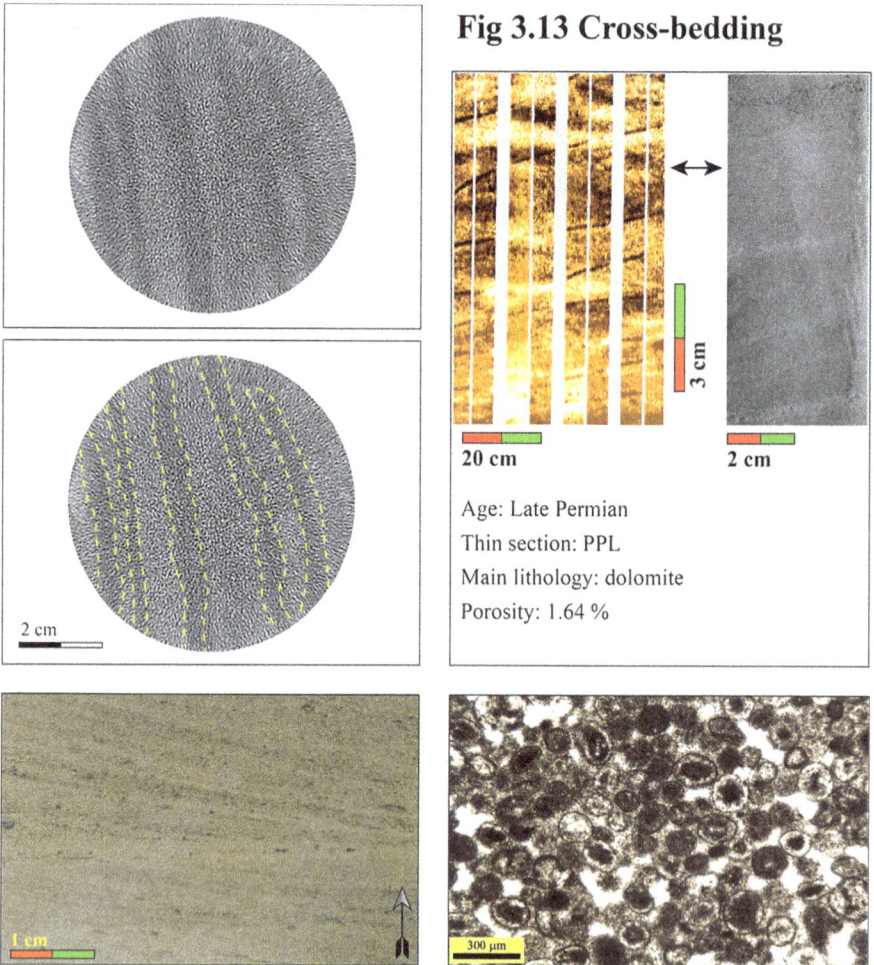

Fig 3.13 Cross-bedding

20 cm

3 cm

2 cm

Age: Late Permian

Thin section: PPL

Main lithology: dolomite

Porosity: 1.64 %

FIGURE 3.13

Beddings are planar structures in sedimentary rocks. They are routinely formed in grain-dominated textures. Some layers are not parallel with the general slope of the stratification (close-up and whole core photos). Each layer has its special characteristics including density (different porosity or mineralogy). In this sample, cross-beddings are marked by dark and light colors in CT images (dashed lines). The grain-dominated texture is obvious in thin section photo. Thick layers are also visible in FMI image.

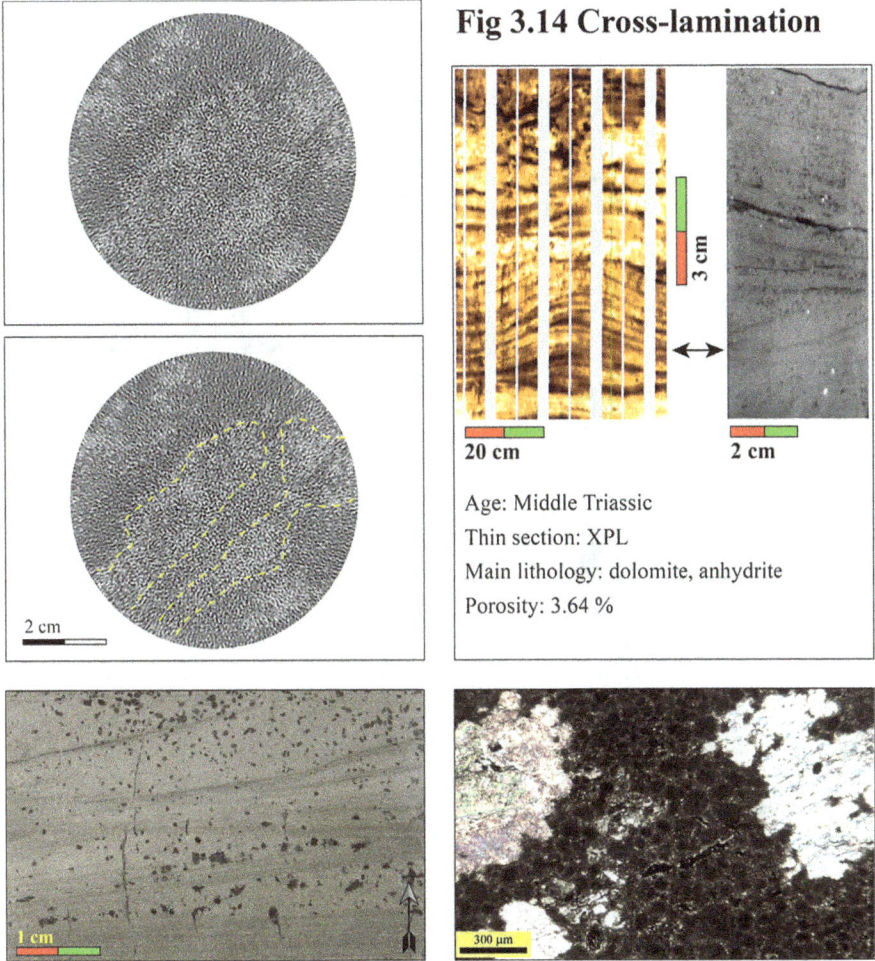

Fig 3.14 Cross-lamination

Age: Middle Triassic

Thin section: XPL

Main lithology: dolomite, anhydrite

Porosity: 3.64 %

FIGURE 3.14

Cross-laminations are evident in this sample. These laminations are seen in FMI image as wavy lines with different angles. They are also visible in whole core, close-up and CT images. In CT slices, they are distinguishable with alternation of dark and light gray color bands (dashed lines). Anhydrite pseudomorphs are frequent in the sample which are characterized with while color in the CT image.

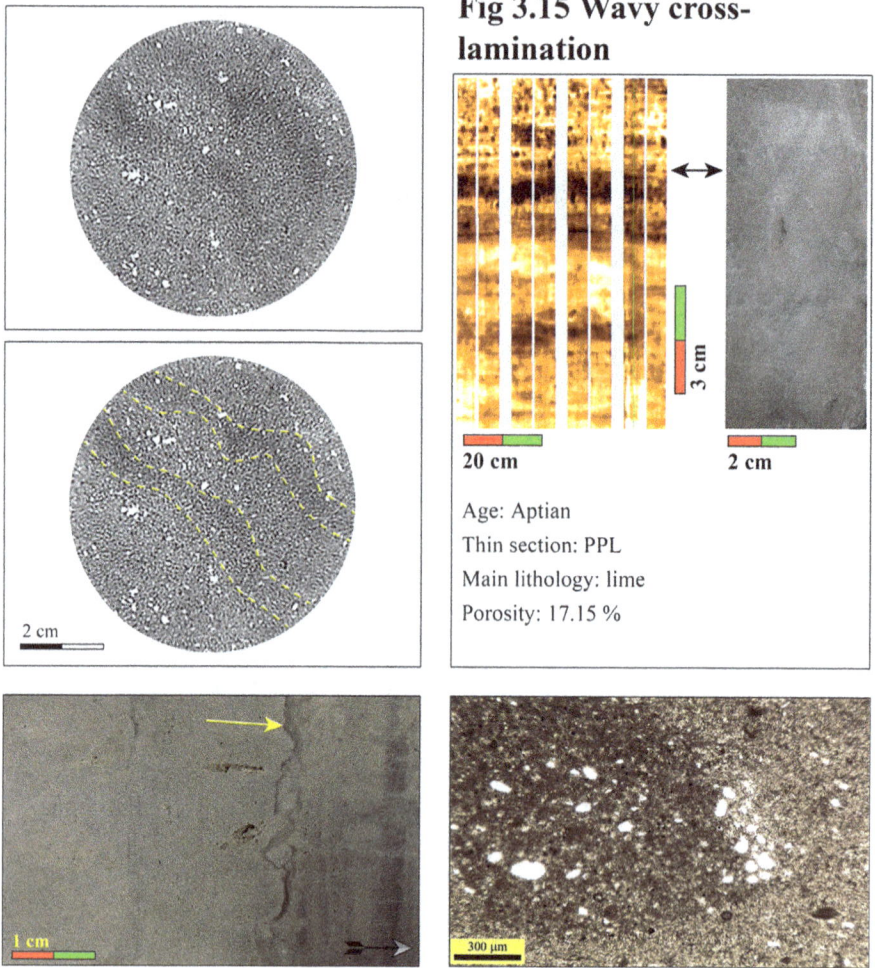

Fig 3.15 Wavy cross-lamination

Age: Aptian
Thin section: PPL
Main lithology: lime
Porosity: 17.15 %

FIGURE 3.15
Wavy laminations are created primarily by oscillating depositional currents or secondary by difference in density of two adjacent layers. Anyway, they are marked by alternation of parallel wavy dark and white lines in CT images (dashed lines) and core photos (yellow arrows). These laminations are also detectable in FMI images. Moldic pores and small anhydrite nodules are also present in the sample.

Stylolites are routinely seen as dark, jagged surfaces with variable widths. Solution seams are also marked as dark lines, but the surface is smoother and usually thicker than stylolites. Both of these structures are dome-like in many cases and so closed curves are also seen, although they are not frequent. Routinely, clays and organic materials remain after dissolution of

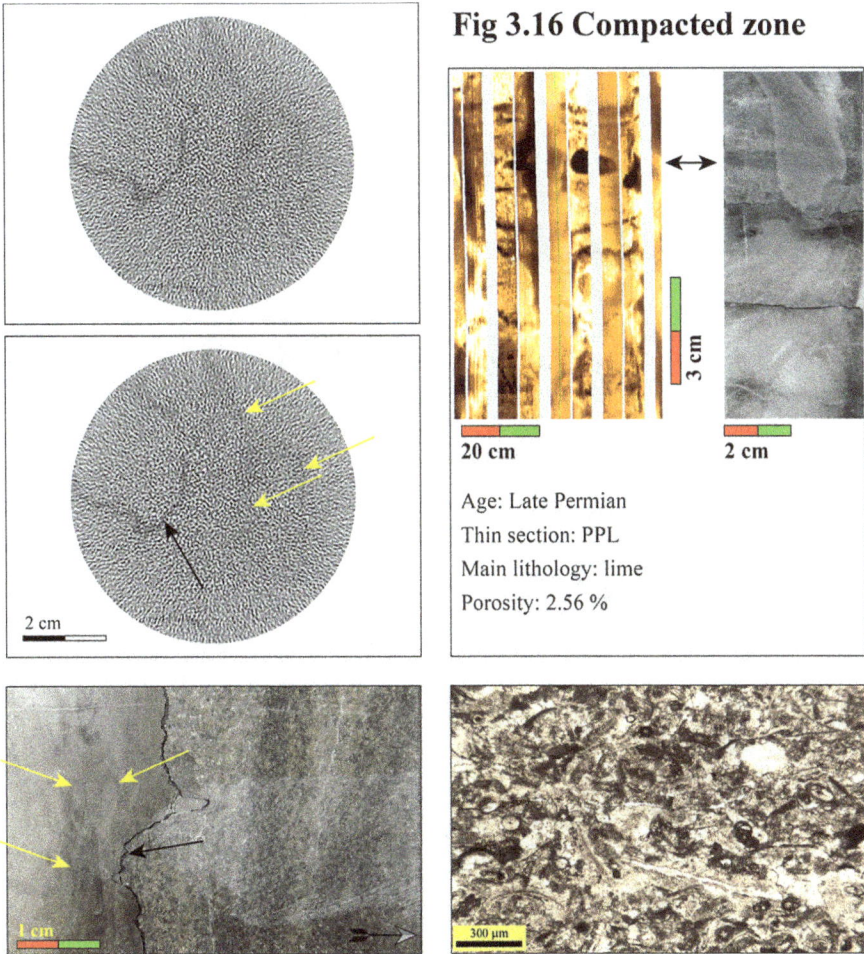

Fig 3.16 Compacted zone

Age: Late Permian
Thin section: PPL
Main lithology: lime
Porosity: 2.56 %

FIGURE 3.16
In addition to the main stylolites (black arrows), there are also some other compacted-related features (yellow arrows) in this sample. They are seen as thin dark veins in CT slice as well as close-up photo (yellow arrows). All of these lines are darker than the background because of lower density. They are composed of clays and organic materials, which are less dense compared to the background minerals.

carbonate minerals. Both of them are less dense than surrounding carbonates and therefore these planar structures are marked by dark surfaces in CT images. Figures 3.16–3.33 of this chapter show the trace of these structures in CT images. They are also detectable in FMI images, thin section studies and core description. All of these proxies are presented along with CT photos.

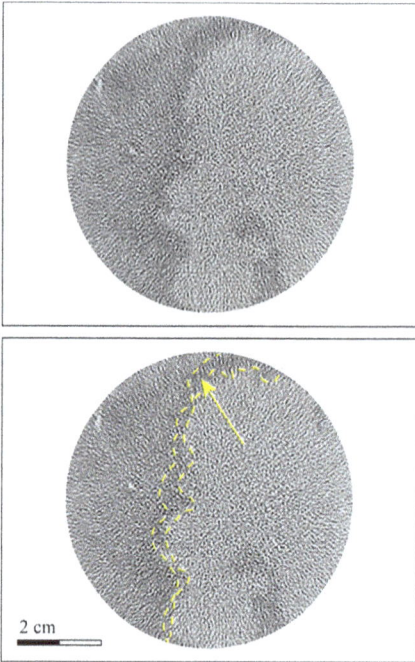

Fig 3.17 Compaction

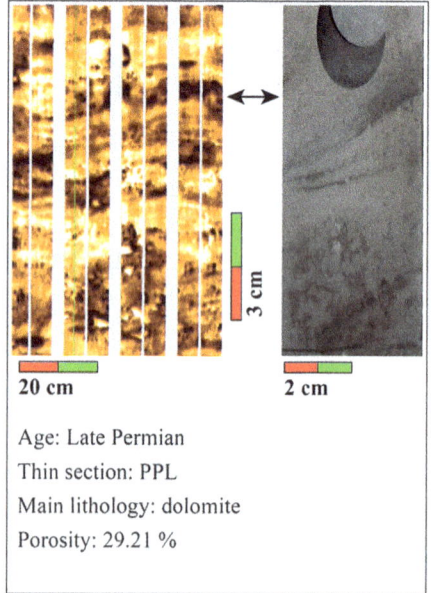

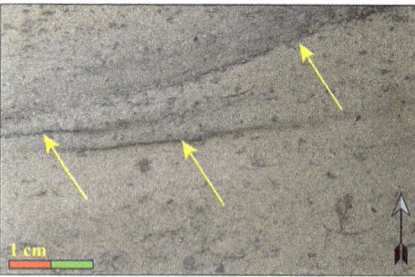

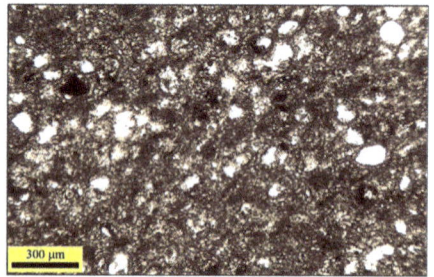

Age: Late Permian
Thin section: PPL
Main lithology: dolomite
Porosity: 29.21 %

FIGURE 3.17

Two layers with different properties are seen in these pictures. These layers are separated by thin veins of insoluble materials. The veins (yellow arrows) show the effect of compaction in this carbonate sample. In fact, they are at the begging stage of stylolite formation, but the pressure and/or insoluble materials were not enough.

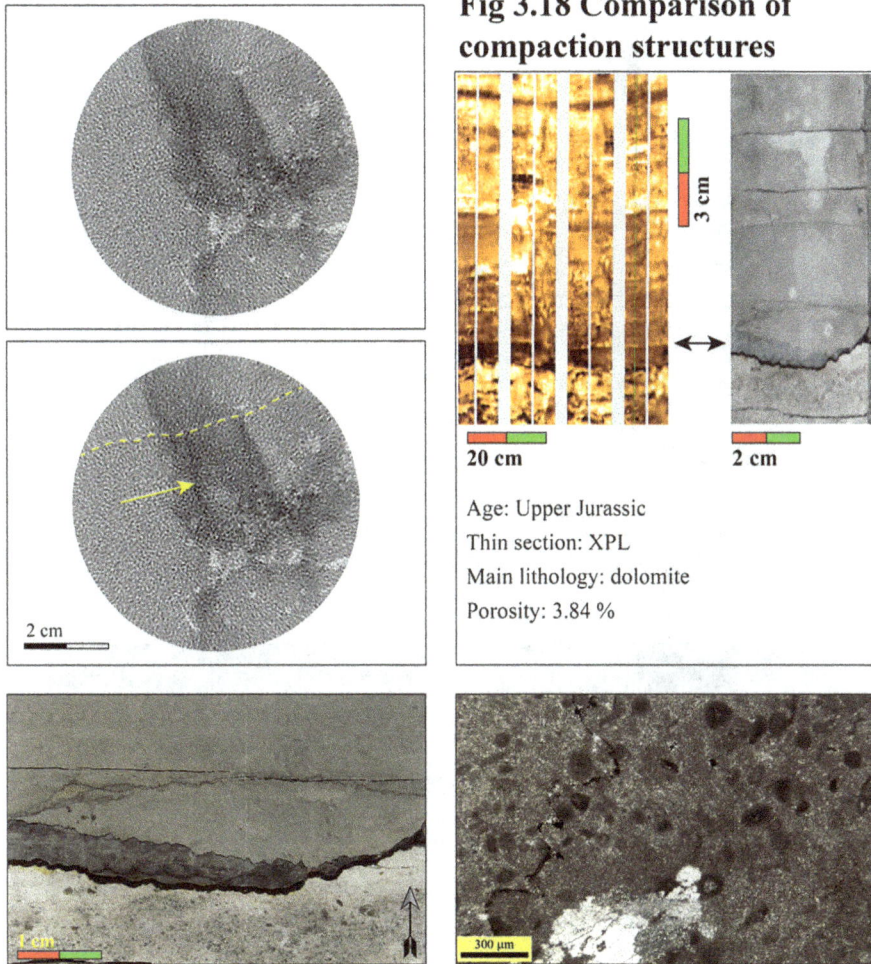

Fig 3.18 Comparison of compaction structures

Age: Upper Jurassic
Thin section: XPL
Main lithology: dolomite
Porosity: 3.84 %

FIGURE 3.18
A stylolite (dashed line) and a compaction zone (yellow arrow) are seen in this CT image. It is clearly seen that stylolite is marked by a dark line while compaction zone is traced with a dark zone. These features are also visible in whole core, close-up and FMI images. With low resistivity, they are marked by dark color in FMI log.

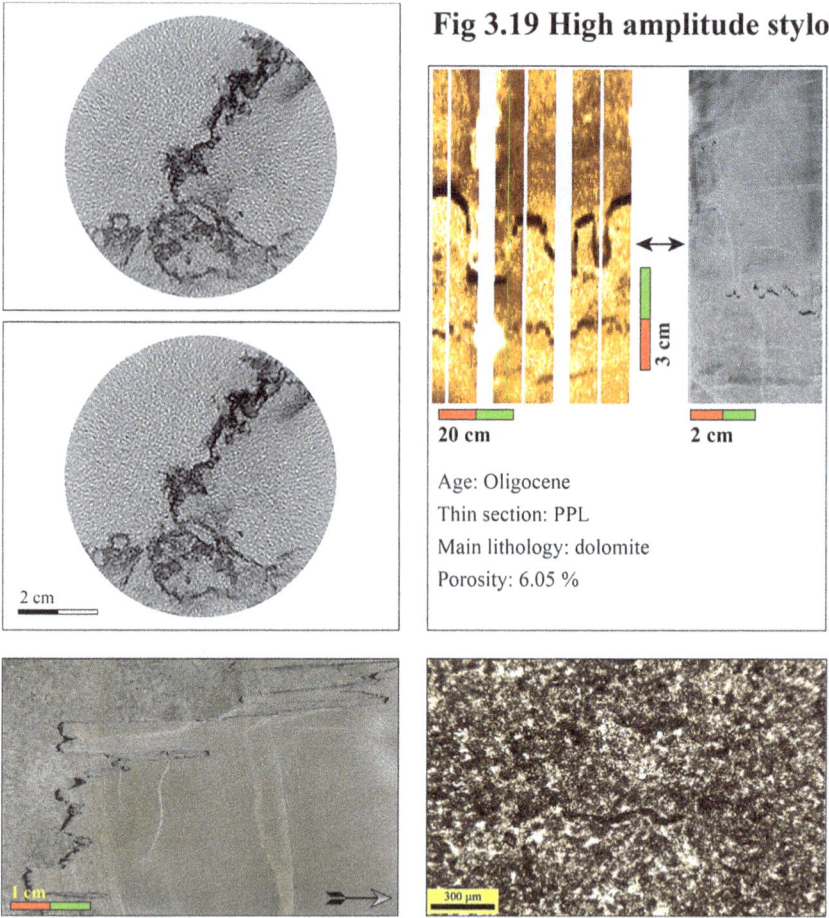

Fig 3.19 High amplitude stylolite

Age: Oligocene
Thin section: PPL
Main lithology: dolomite
Porosity: 6.05 %

FIGURE 3.19
High-amplitude stylolites are clearly seen in CT image because considerable part of the stylolite is intersected by the CT slice. This stylolite is visible in CT, FMI, close-up and whole core photos. Facies changes at the stylolite boundary but densities are the same in two facies and so the CT image is homogenous.

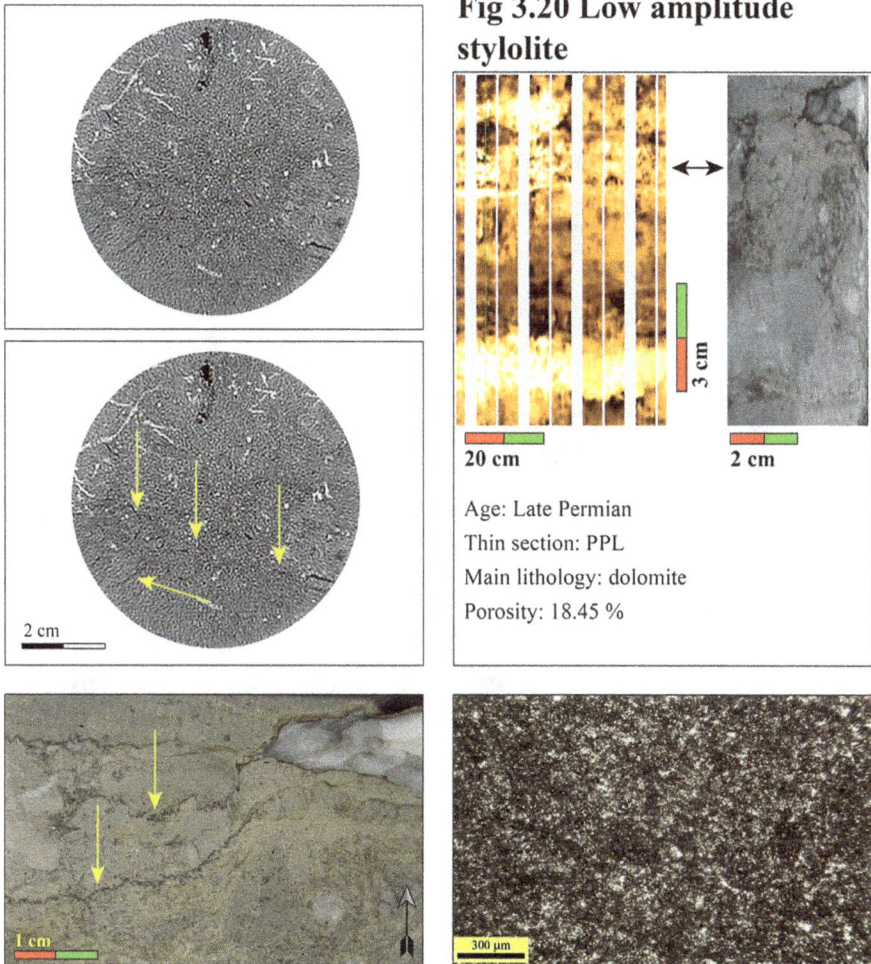

Fig 3.20 Low amplitude stylolite

20 cm

2 cm

Age: Late Permian

Thin section: PPL

Main lithology: dolomite

Porosity: 18.45 %

FIGURE 3.20

Stylolites have different amplitudes in carbonate rocks. High-amplitude stylolites are easier to recognize. Here, a low-amplitude stylolite is observed. Distinguishing of these planar structures is not easy in CT images. They are marked by nearly straight, dark lines in these photos (arrows). Anhydrite is frequent in the sample as replacements.

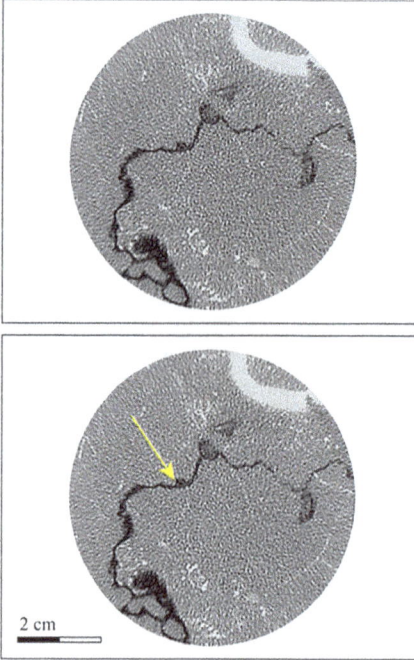

Fig 3.21 Solution seam

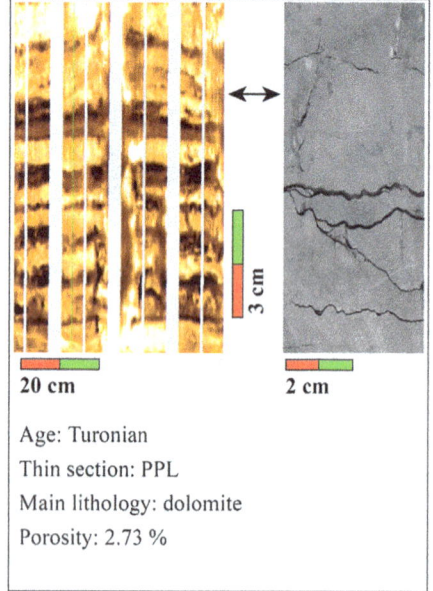

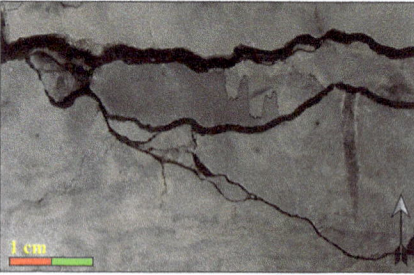

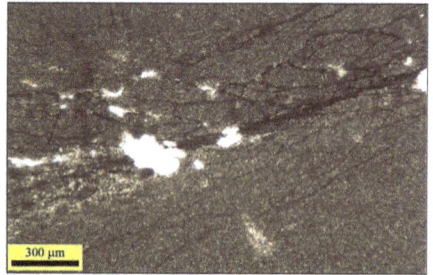

Age: Turonian

Thin section: PPL

Main lithology: dolomite

Porosity: 2.73 %

FIGURE 3.21

Thick, linear accumulations of insoluble materials are called solution seams. These materials are routinely less dense than the rock matrix. Therefore, they are marked by dark lines in CT images (yellow arrow). Here, a thick solution seam is obvious in mudstone texture. As it is nearly horizontal, its major part is visible in the CT image.

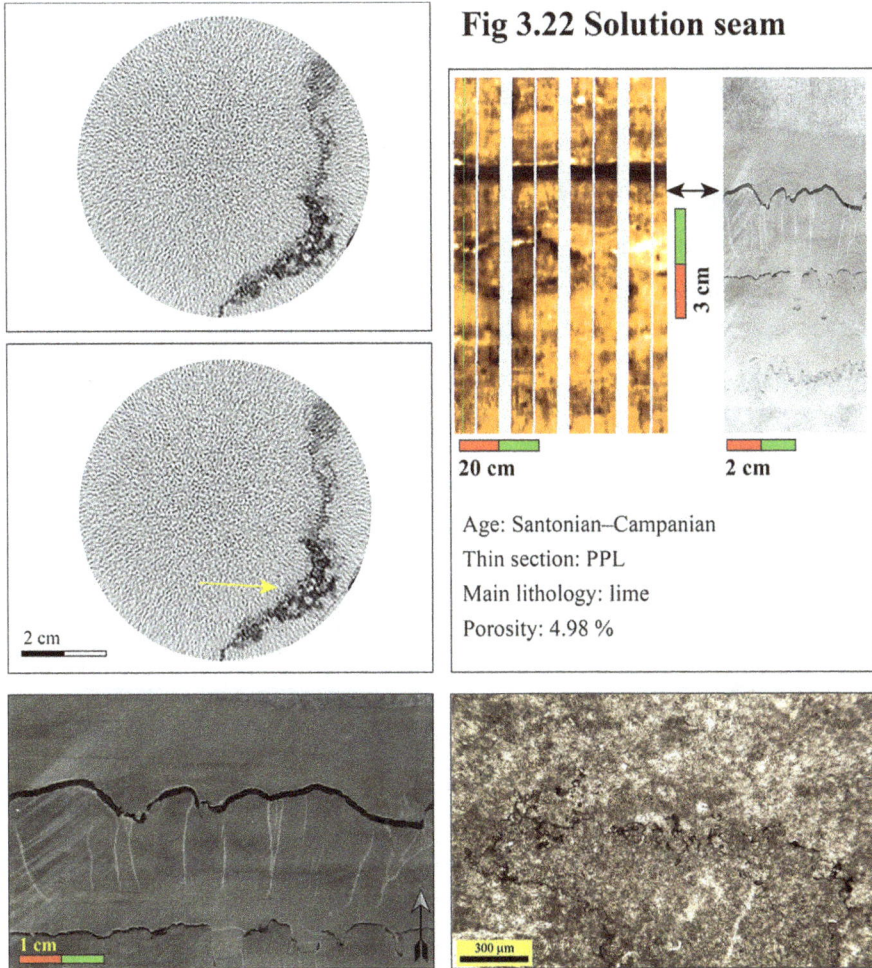

Fig 3.22 Solution seam

20 cm   2 cm

Age: Santonian–Campanian
Thin section: PPL
Main lithology: lime
Porosity: 4.98 %

FIGURE 3.22

Stylolites and solution seams are two frequent compaction-related structures in carbonate reservoirs. Thick solution seams are easily recognizable on transverse CT scan images (arrow). In comparison with stylolites, they are thicker and have a smooth surface. Carbonates are dissolved and other insoluble minerals, commonly clays, form these surfaces. Clay minerals have lower density and so these structures are distinguishable with their darker color in CT images.

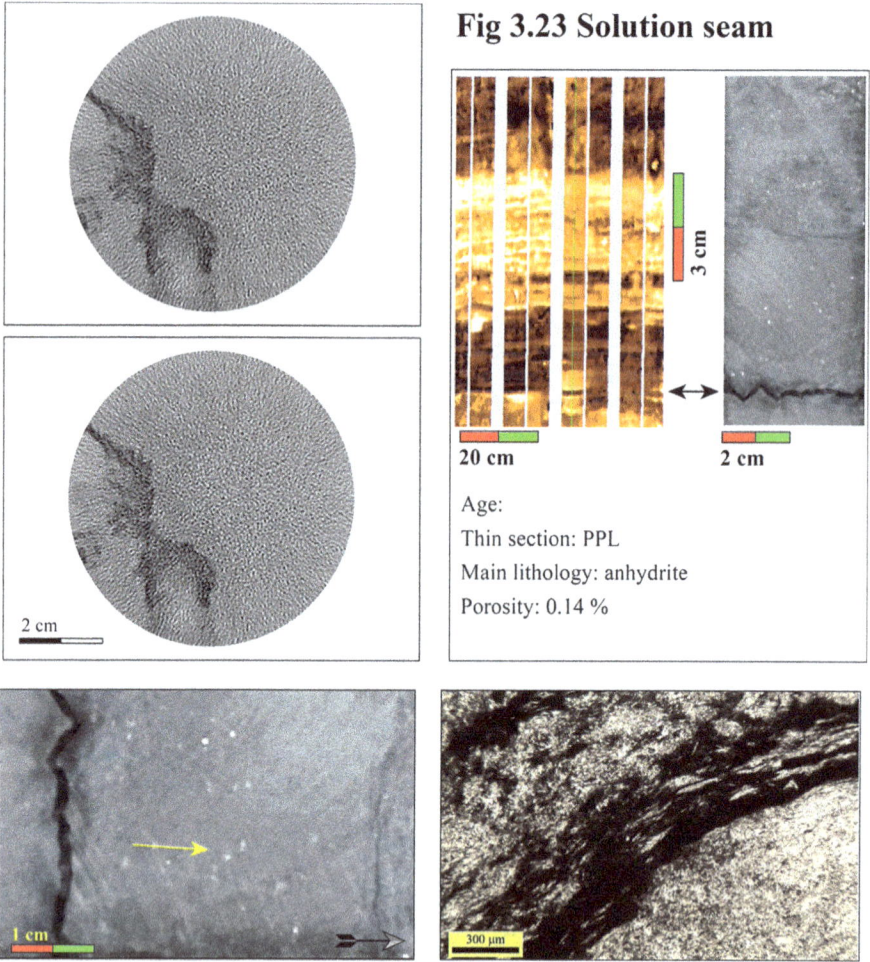

Fig 3.23 Solution seam

Age:
Thin section: PPL
Main lithology: anhydrite
Porosity: 0.14 %

FIGURE 3.23

Solution seams are developed in all lithologies. These images show a solution seam in anhydrite lithology. They are routinely formed due to the high organic content as well as clay minerals in evaporites. They can be recognized by black color in CT (due to low density), whole core, close-up, thin section and FMI (due to low resistivity) images.

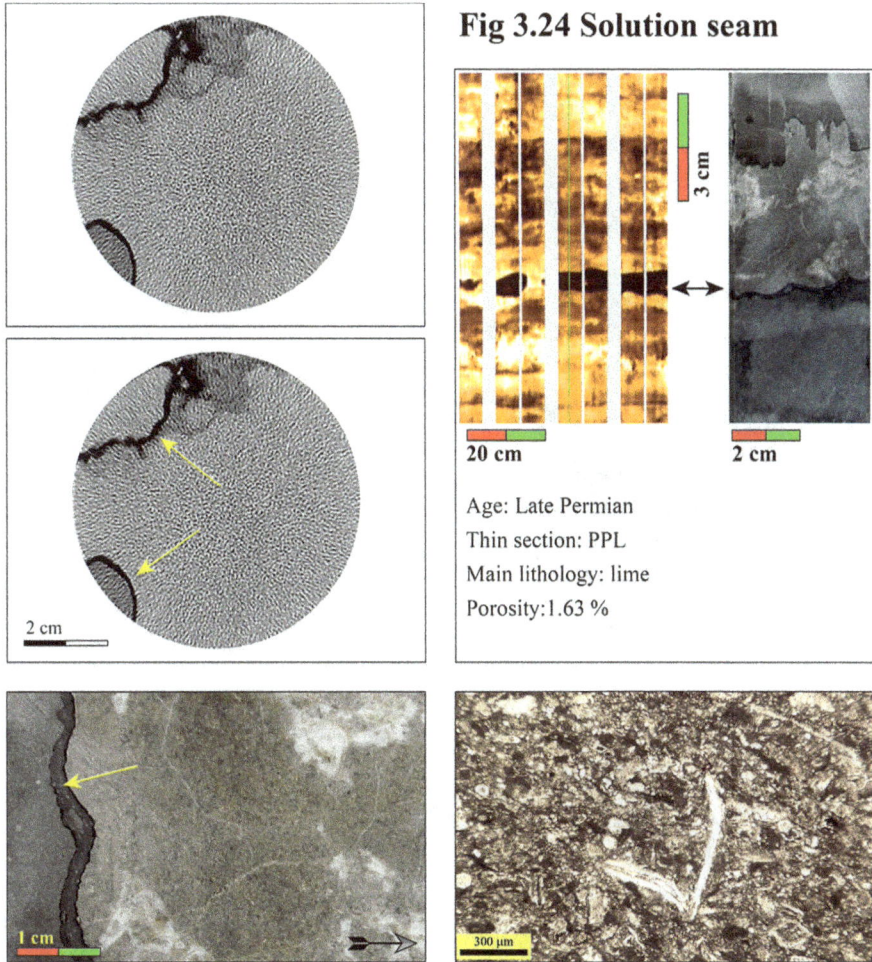

Fig 3.24 Solution seam

Age: Late Permian
Thin section: PPL
Main lithology: lime
Porosity: 1.63 %

FIGURE 3.24

Solution seams are thicker and smoother than stylolites. These features are easily distinguishable in CT images. As these are dome-like surfaces in many cases, the intersection is viewed as open or closed curvatures. Such as stylolites, these planar structures are rich in low-density materials, such as clays or organic materials. Therefore, they are dark gray in CT images.

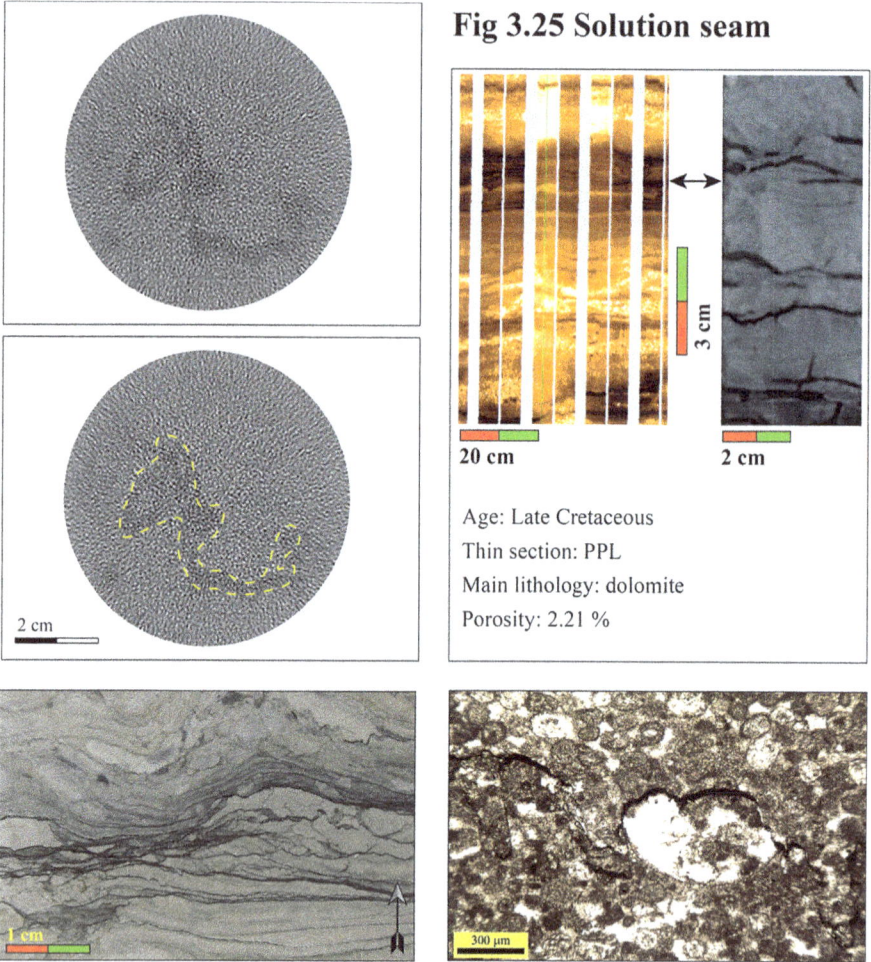

Fig 3.25 Solution seam

Age: Late Cretaceous
Thin section: PPL
Main lithology: dolomite
Porosity: 2.21 %

FIGURE 3.25

A solution seam is obvious in CT scan image of a core sample with dolomite lithology. Solution seams are thicker than stylolites and so are more obvious in CT images. They could be detected on FMI results, whole core and close-up photos. Presence of clay minerals and organic matter leads to lower density in comparison with carbonates, especially dolomite. Stylolites and solution seams are planar structures; therefore, they are seen as curved lines in transversal CT images.

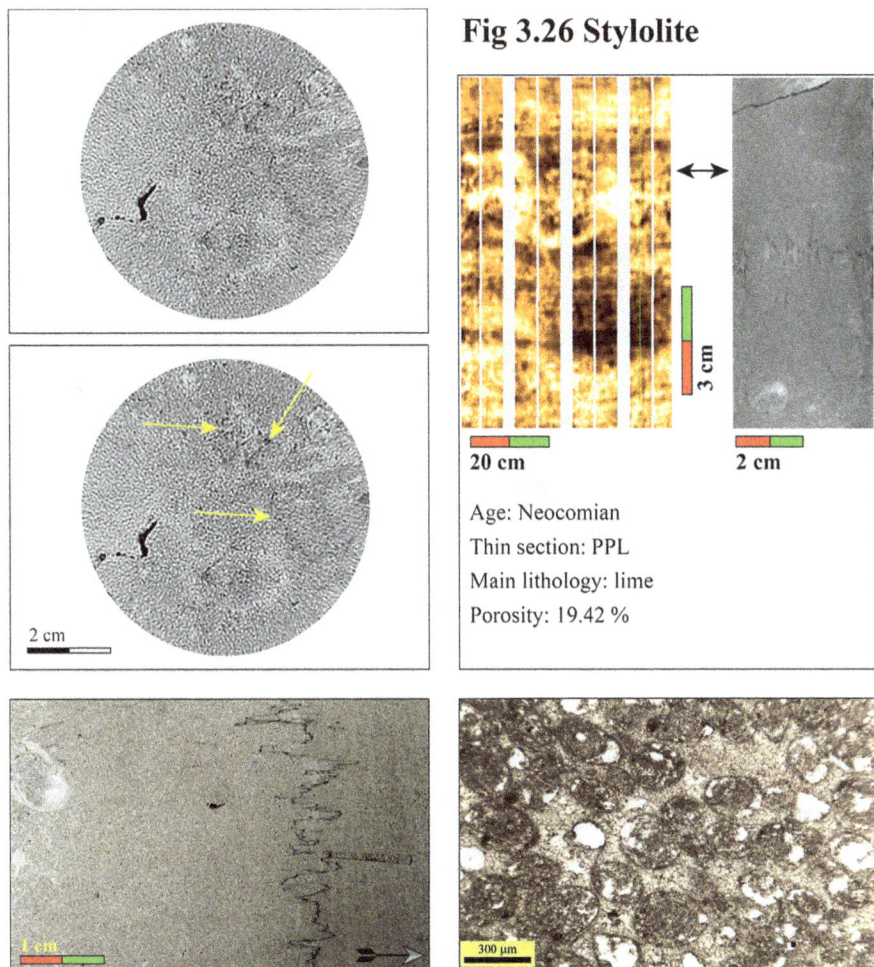

Fig 3.26 Stylolite

Age: Neocomian
Thin section: PPL
Main lithology: lime
Porosity: 19.42 %

FIGURE 3.26

A high-amplitude stylolite in a grain-dominated carbonate facies is visible in this image. It is marked by a jagged dark gray line in CT image (arrows). The sample has high moldic porosity. Therefore, the CT image is generally dark. Anhydrites have been precipitated around stylolite surface in the direction of minimum stress.

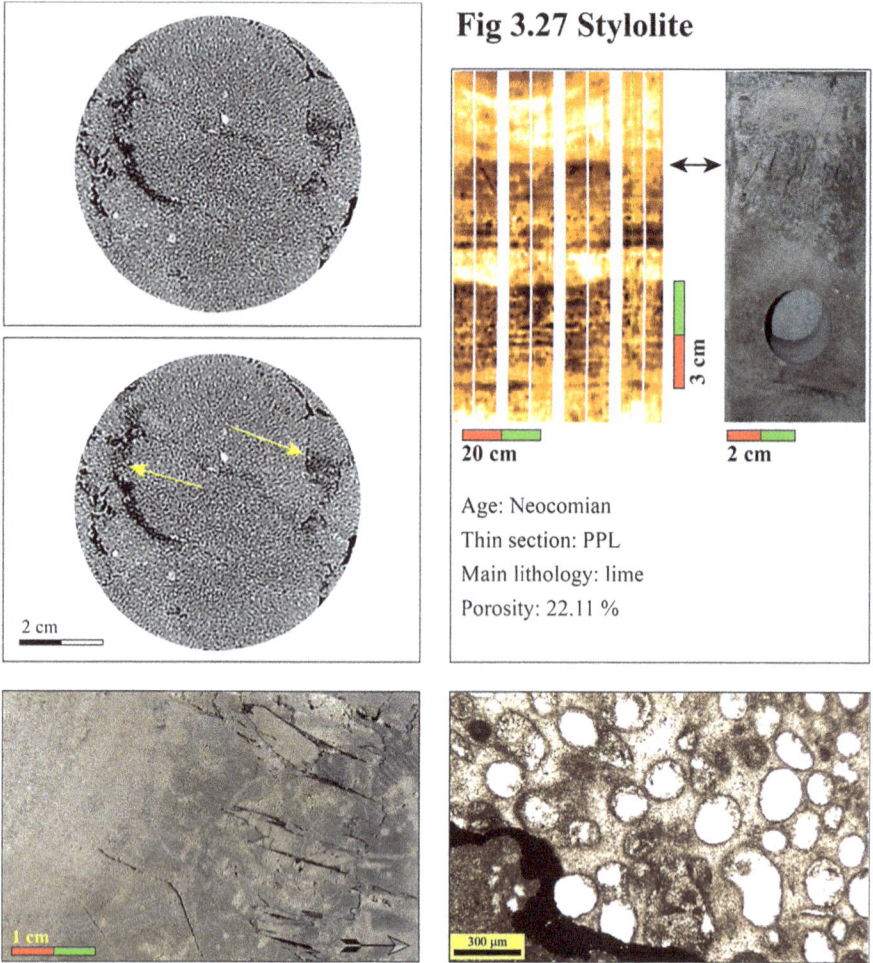

Fig 3.27 Stylolite

20 cm   2 cm

Age: Neocomian
Thin section: PPL
Main lithology: lime
Porosity: 22.11 %

FIGURE 3.27
A high-amplitude, thick stylolite is obvious in this sample. Insoluble materials such as clays or organic materials are accumulated in these surfaces. The density of these materials is lower than the rock and so they are seen as black lines in CT images (arrows). Various parts of the stylolite surface have been intersected by the CT slice and so the resulted dark line can be seen in various parts of the image.

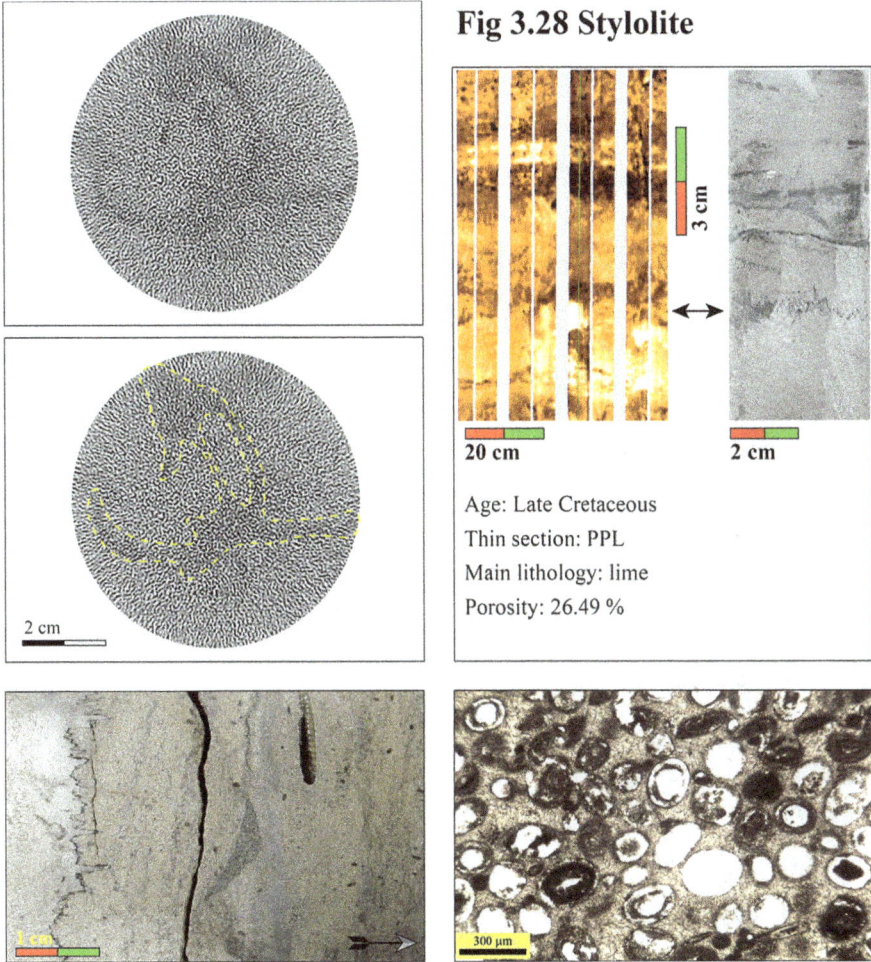

Fig 3.28 Stylolite

Age: Late Cretaceous
Thin section: PPL
Main lithology: lime
Porosity: 26.49 %

FIGURE 3.28

Depending on the position of the CT slice, various parts of a stylolite can be seen in the image. It also depends on the amplitude and frequency of the stylolite surface. In this image, a large part of the stylolite is seen (dashed line). Large stylolites are also visible in FMI image. The sample has high moldic porosity which reduces the stylolite and background contrast in CT slice.

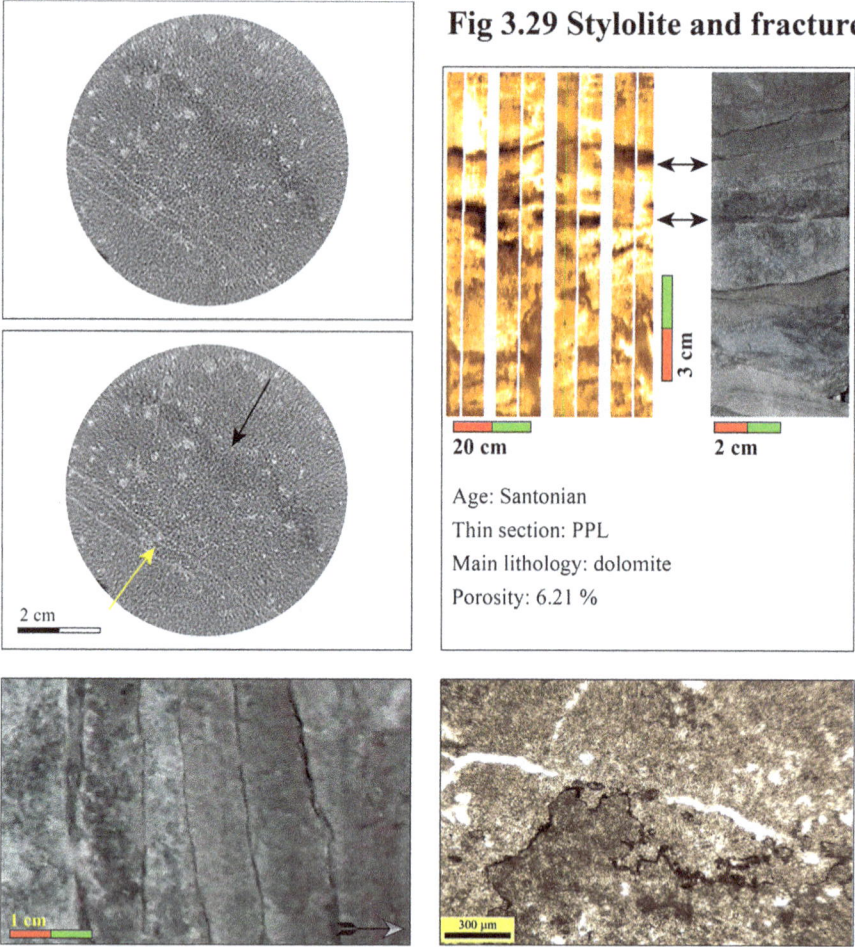

Fig 3.29 Stylolite and fracture

Age: Santonian
Thin section: PPL
Main lithology: dolomite
Porosity: 6.21 %

FIGURE 3.29

Overburden pressure creates horizontal stylolites. Fractures are routinely filled by a dense material (anhydrite in this case) while stylolites are formed from clay minerals, and therefore, they can be recognized by their light (yellow arrow) and dark (black arrow) colors, respectively. Both of them can be marked in thin section studies in some cases.

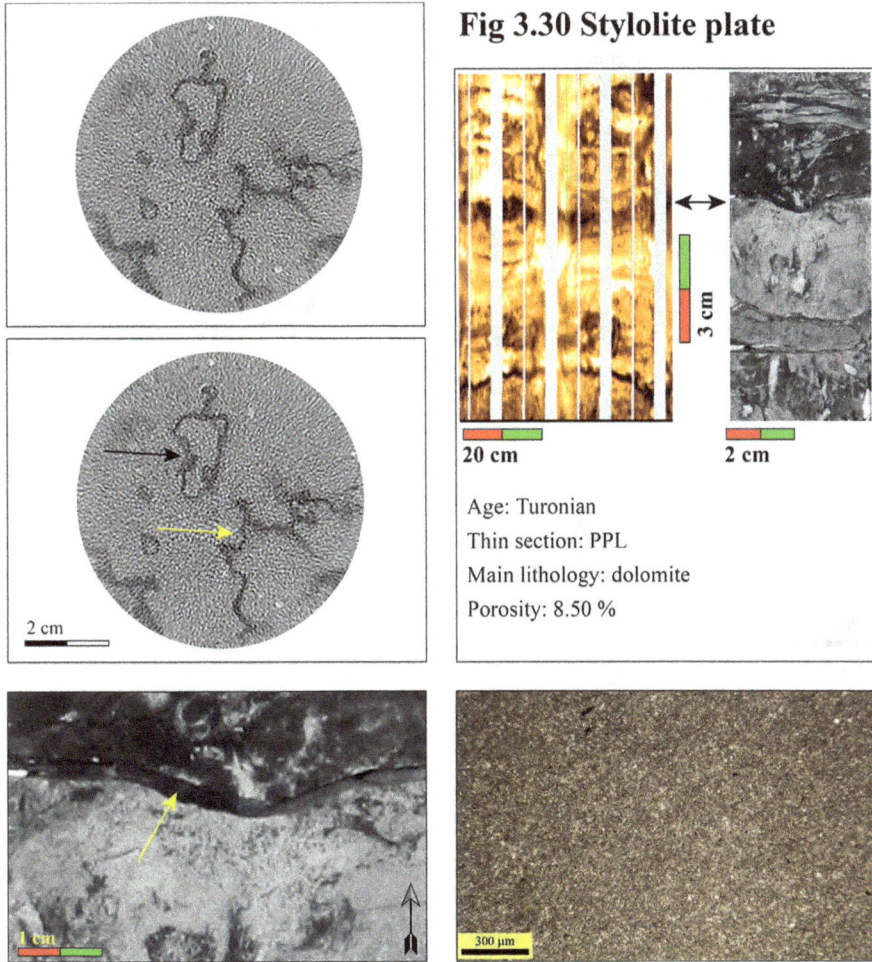

Fig 3.30 Stylolite plate

Age: Turonian
Thin section: PPL
Main lithology: dolomite
Porosity: 8.50 %

FIGURE 3.30

Stylolites are routinely seen as jagged curved surfaces (yellow arrow) on CT images but closed and dark areas (black arrow) are also visible. This is due to the curvature of the stylolite surface (dome-like shape). Clay minerals and organic matters are frequent in these surfaces with low density and so they are recognized by dark lines in CT images. They form dark lines on FMI images, too.

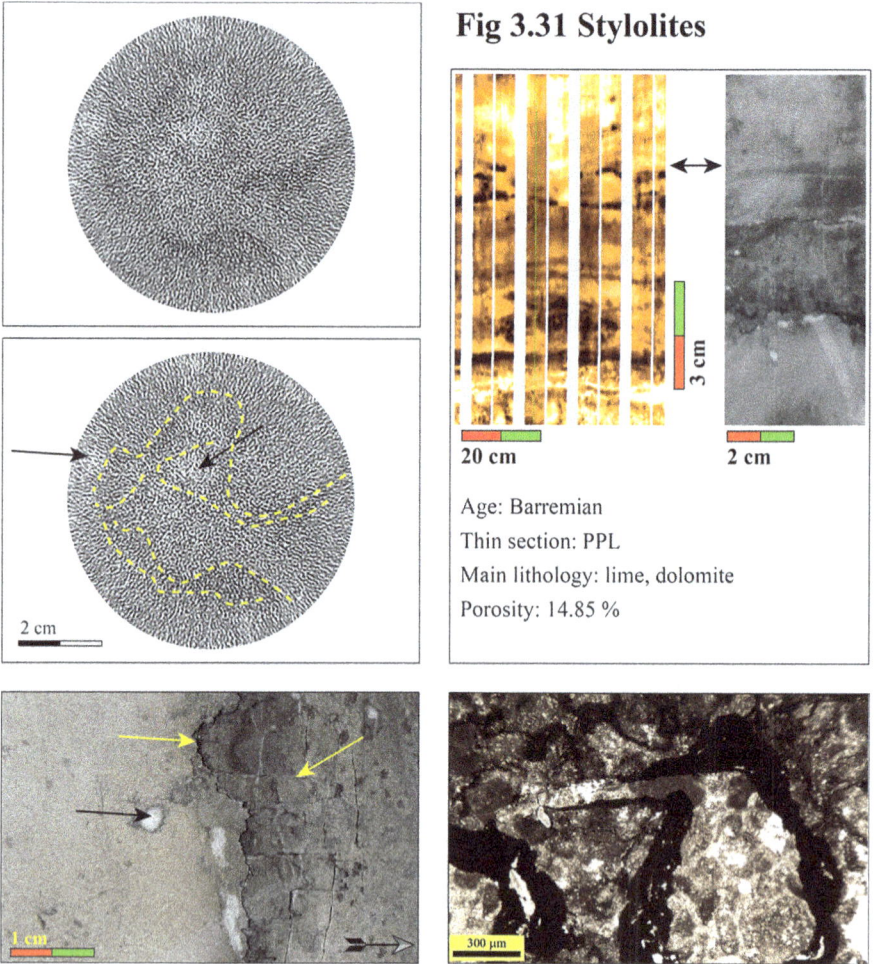

Fig 3.31 Stylolites

Age: Barremian
Thin section: PPL
Main lithology: lime, dolomite
Porosity: 14.85 %

FIGURE 3.31

A combination of various types of stylolites is seen in this sample. Stylolites are marked by dark gray color in CT image (dashed lines). They are also visible in whole core and close-up (yellow arrows) photos. Anhydrite nodules are present in the sample (black arrows). Thin section studies show that these stylolites are rich in organic materials.

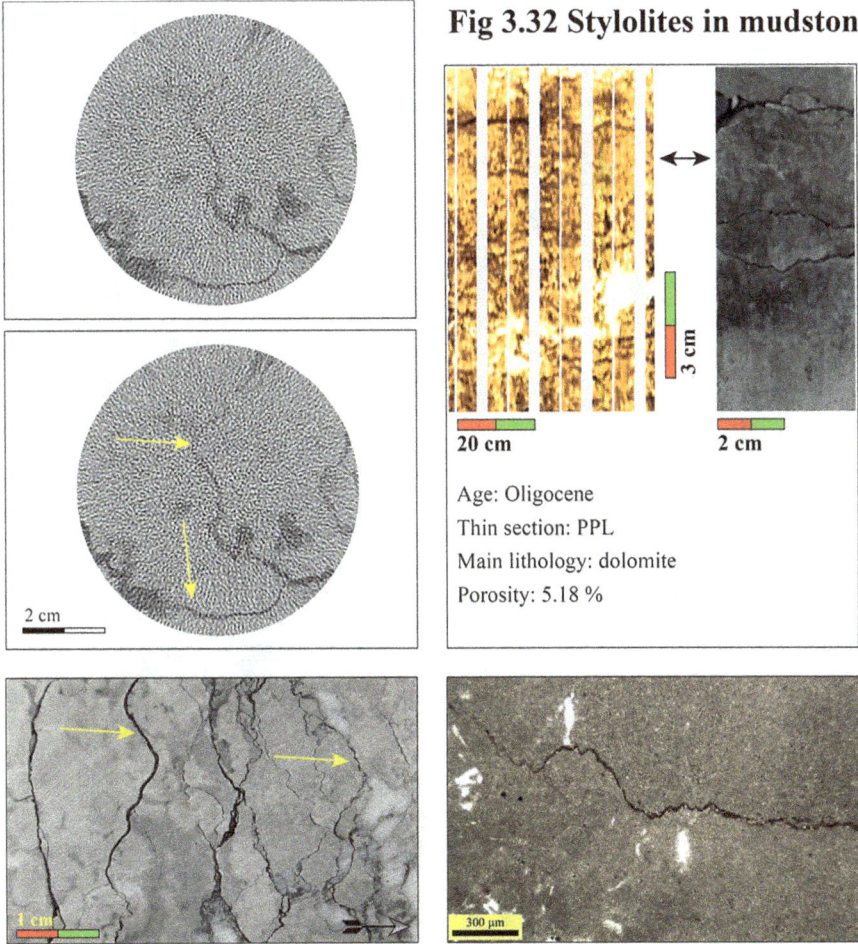

Fig 3.32 Stylolites in mudstone

Age: Oligocene
Thin section: PPL
Main lithology: dolomite
Porosity: 5.18 %

FIGURE 3.32

Mudstones are susceptible to stylolite formation as they contain high organic and clay contents. These structures are observable as dark gray lines in CT images because they contain organic materials or clay minerals with low density. These stylolites are obvious on whole core, close-up and FMI images as well as thin section studies (arrows).

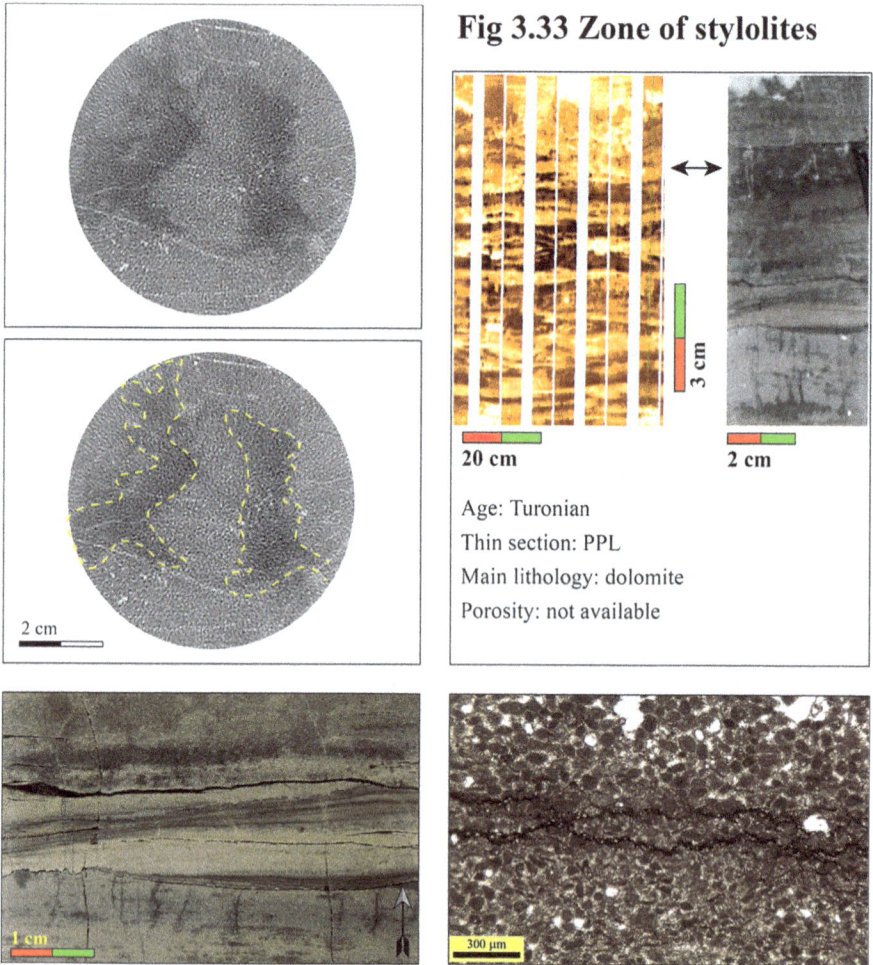

Fig 3.33 Zone of stylolites

20 cm   2 cm

Age: Turonian
Thin section: PPL
Main lithology: dolomite
Porosity: not available

FIGURE 3.33
Many thin stylolites are visible in this sample. Therefore, a zone with dark color has been created in the CT slice (dashed line). In fact, there is a zone enriched in low-density materials (mainly clays and organic content). These stylolites are also visible in FMI image because they have low resistivity (clay bound waters).

The main part of this chapter considers the effect of various types of fractures in CT images. These include desiccation cracks, hairline and large fractures, both vertical and horizontal fractures, fractured zones, natural hydraulic fracturing, filled and open fractures, induced fractures and tension gashes (Figures 3.34–3.60). Depending on the type of fracture and filling

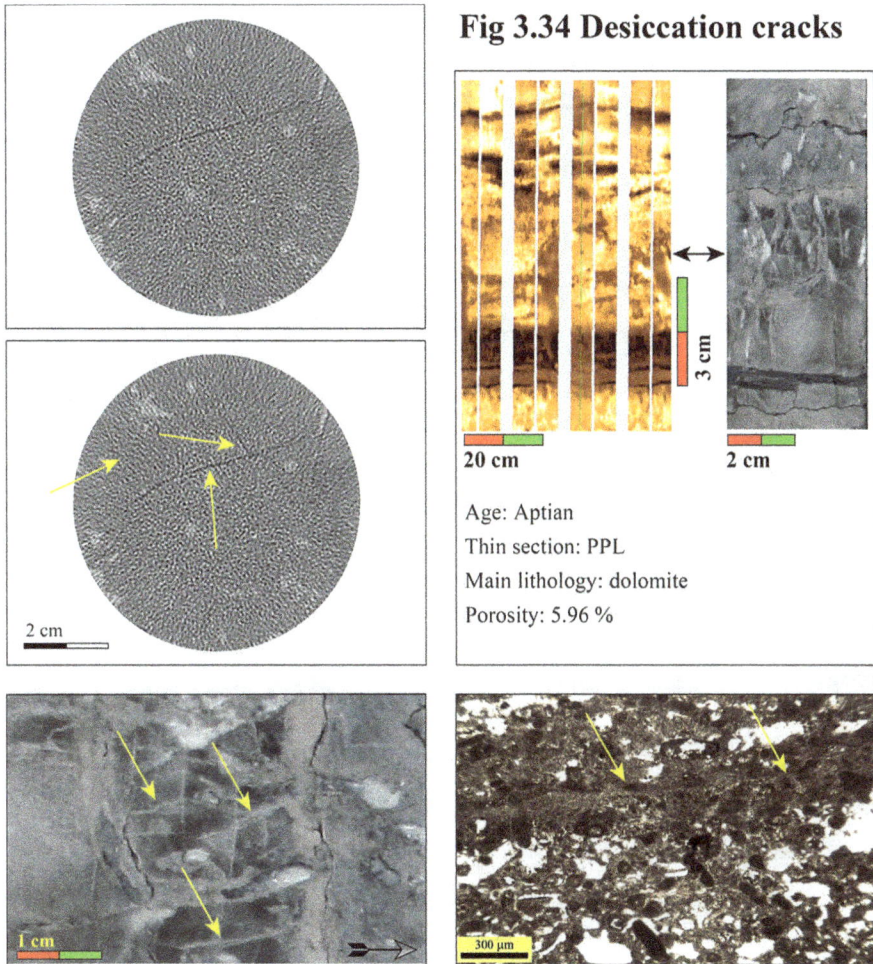

Fig 3.34 Desiccation cracks

Age: Aptian
Thin section: PPL
Main lithology: dolomite
Porosity: 5.96 %

FIGURE 3.34

Desiccation cracks are formed due to dewatering of sediments. Therefore, they indicate a shallow depositional setting. They are created in different directions in mud-dominated (mudstone to wackestone) facies. They usually terminate against each other. These cracks are small and nearly perpendicular to layering and so can be distinguished from fractures in carbonate sediments.

materials, they can be seen as dark or light lines with different characteristics. The last two pages of this chapter show how crushed zones are seen in CT images. Altogether, these pages show the effect of various types of planar structures in CT scan slices. Finally, figures depicting crushed cores are presented (Figures 3.61 and 3.62).

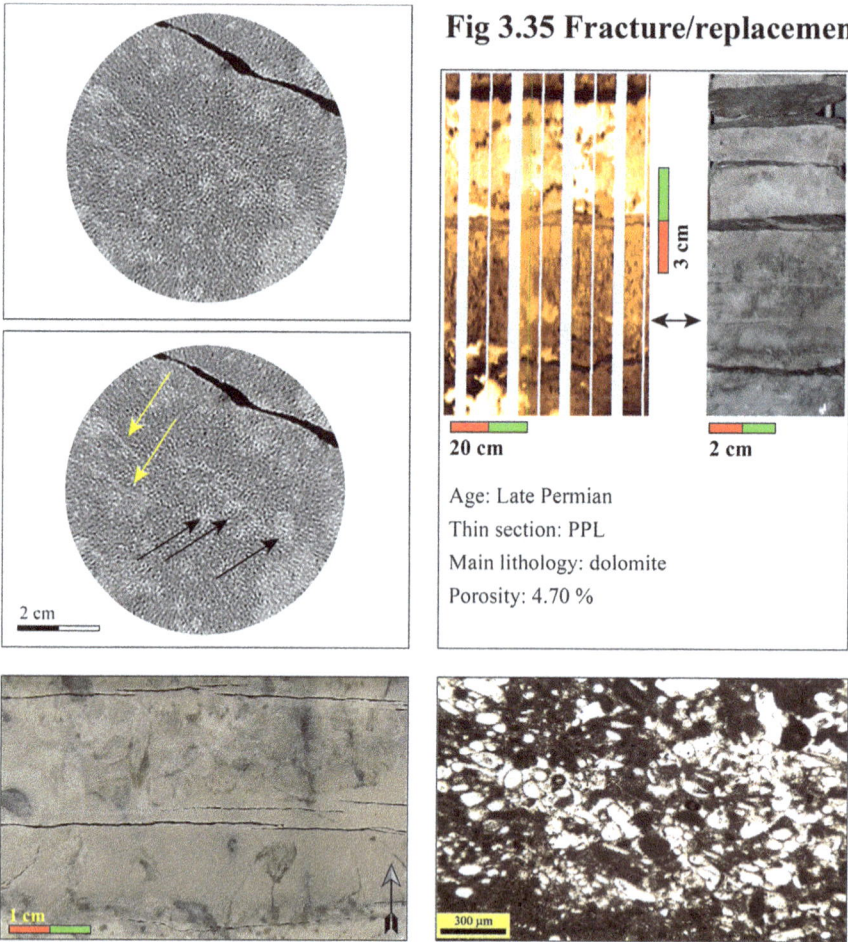

Fig 3.35 Fracture/replacement

Age: Late Permian
Thin section: PPL
Main lithology: dolomite
Porosity: 4.70 %

FIGURE 3.35

Pervasive anhydrite cement has filled both fractures (yellow arrows) and dissolved bioclasts (black arrows). The original shapes of some bioclasts are still visible. Some ooids have been dissolved and filled by anhydrite, too. The core has been broken in coring process which is characterized by a dark thick and smooth surface in CT image.

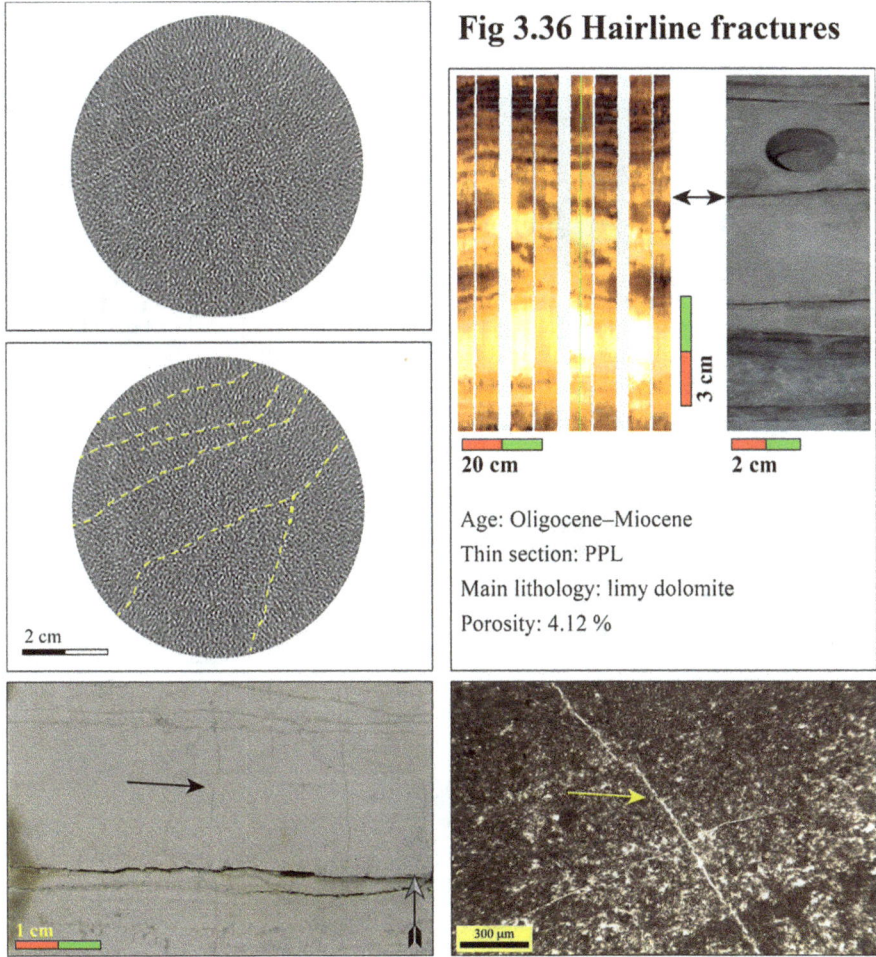

Fig 3.36 Hairline fractures

Age: Oligocene–Miocene

Thin section: PPL

Main lithology: limy dolomite

Porosity: 4.12 %

FIGURE 3.36

Vertical hairline fractures (dashed lines) are seen in mudstone facies. These fractures are not visible in FMI image or whole core photo but are obvious in close-up images and thin section studies (arrows). They have been filled by anhydrite cement, which is a routine diagenetic process in carbonate-evaporite sequences. Anhydrite cement with higher density could be recognized by the milky white color in transversal CT images.

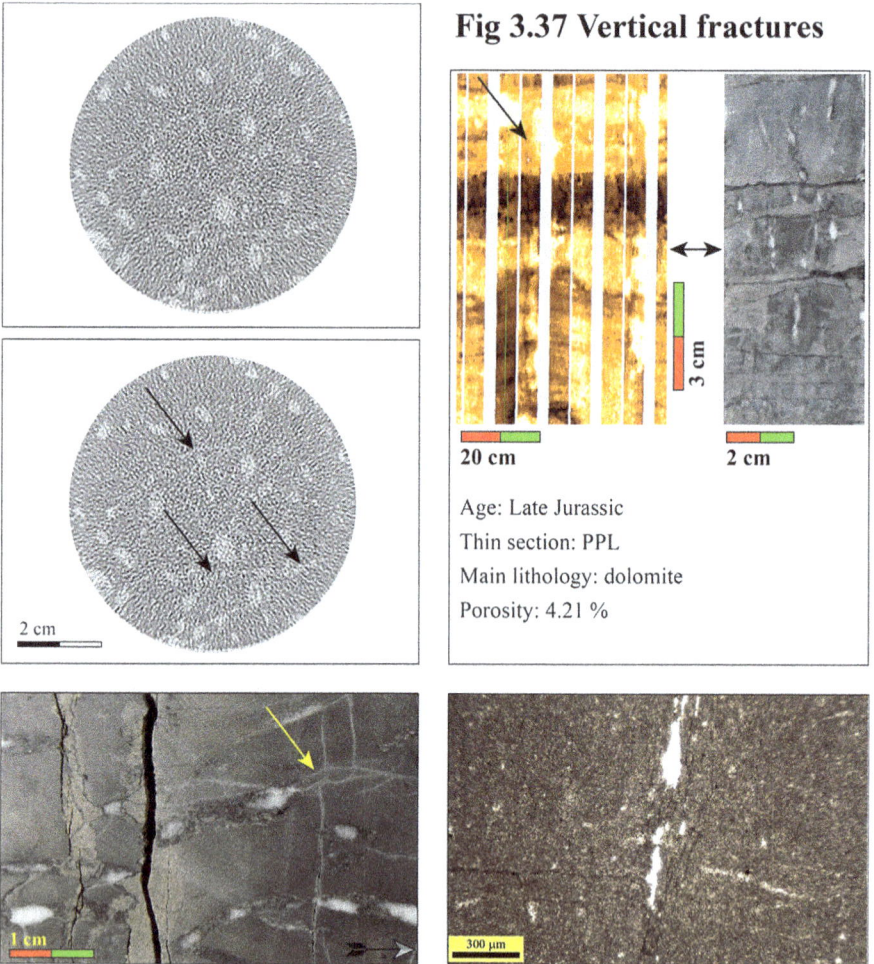

Fig 3.37 Vertical fractures

Age: Late Jurassic
Thin section: PPL
Main lithology: dolomite
Porosity: 4.21 %

FIGURE 3.37

A set of vertical fractures (arrows) are visible in these images. These fractures have been filled with anhydrite and so they are recognized by light color in CT scan images. They are also visible on FMI and whole cores. There are many borings in the sample, which have been filled by anhydrite. Therefore, they are seen as light circles on CT image.

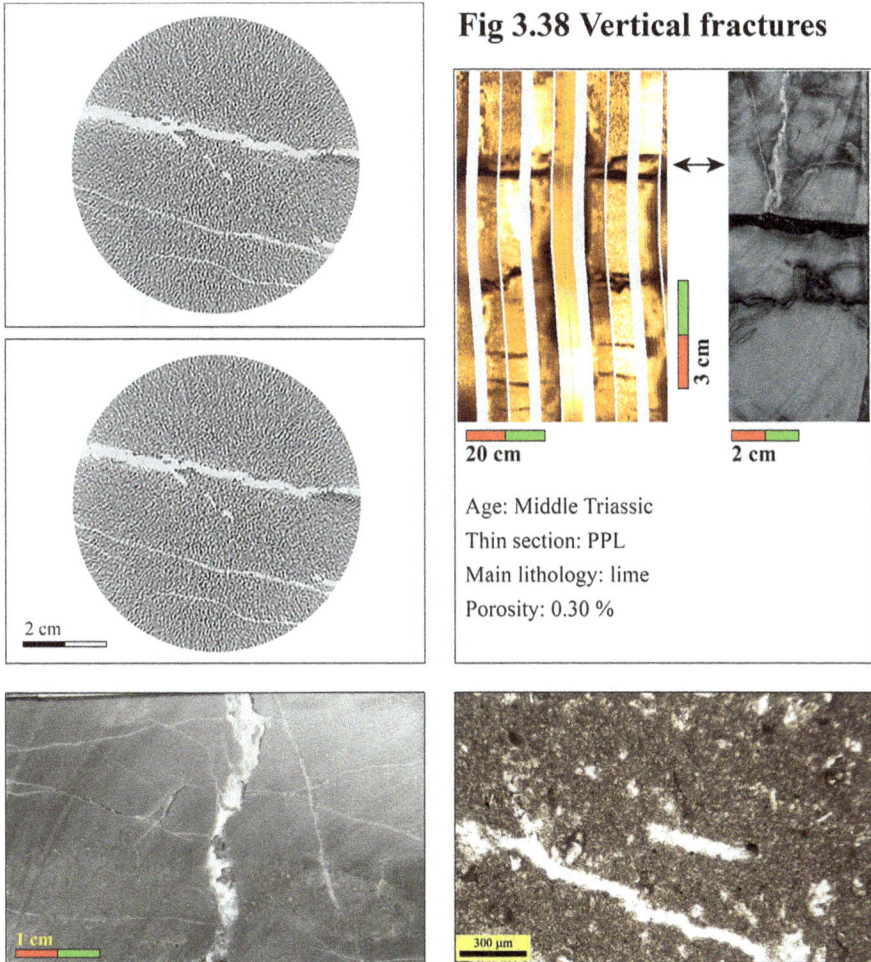

Fig 3.38 Vertical fractures

Age: Middle Triassic
Thin section: PPL
Main lithology: lime
Porosity: 0.30 %

FIGURE 3.38
Vertical fractures have been filled with anhydrite in this sample. They are marked by milky white color in CT image. Large fractures are visible in CT, whole core, FMI and close-up photos, while small fractures are not visible in FMI log. Some of the fractures are not visible in CT slices, too. They are horizontal or nearly horizontal and so are not crossed by the CT plane.

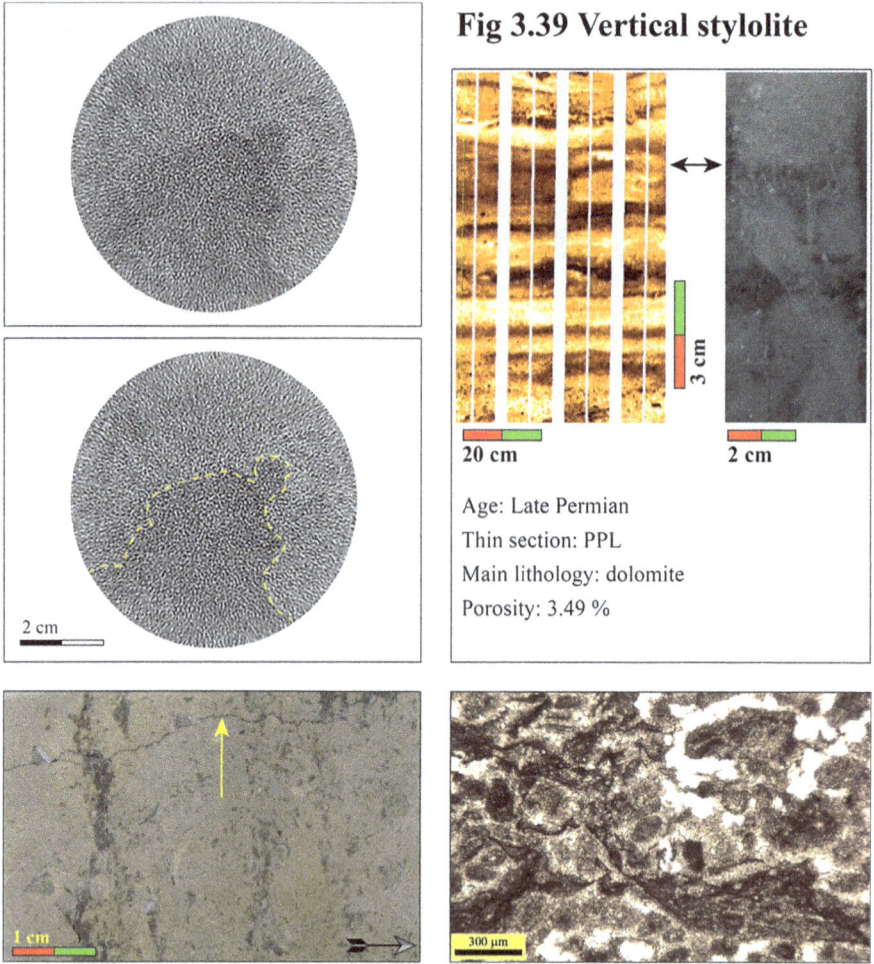

Fig 3.39 Vertical stylolite

20 cm   2 cm

Age: Late Permian
Thin section: PPL
Main lithology: dolomite
Porosity: 3.49 %

FIGURE 3.39
Vertical stylolites are not frequent in carbonate reservoirs because the main stress is vertical in many cases (overburden pressure). Therefore, stylolites are developed parallel to bedding. Here, a vertical stylolite is visible in CT (dashed line) and close-up (arrow) photos. It is not clear in FMI image because the tool resolution is lower than the thickness of this stylolite.

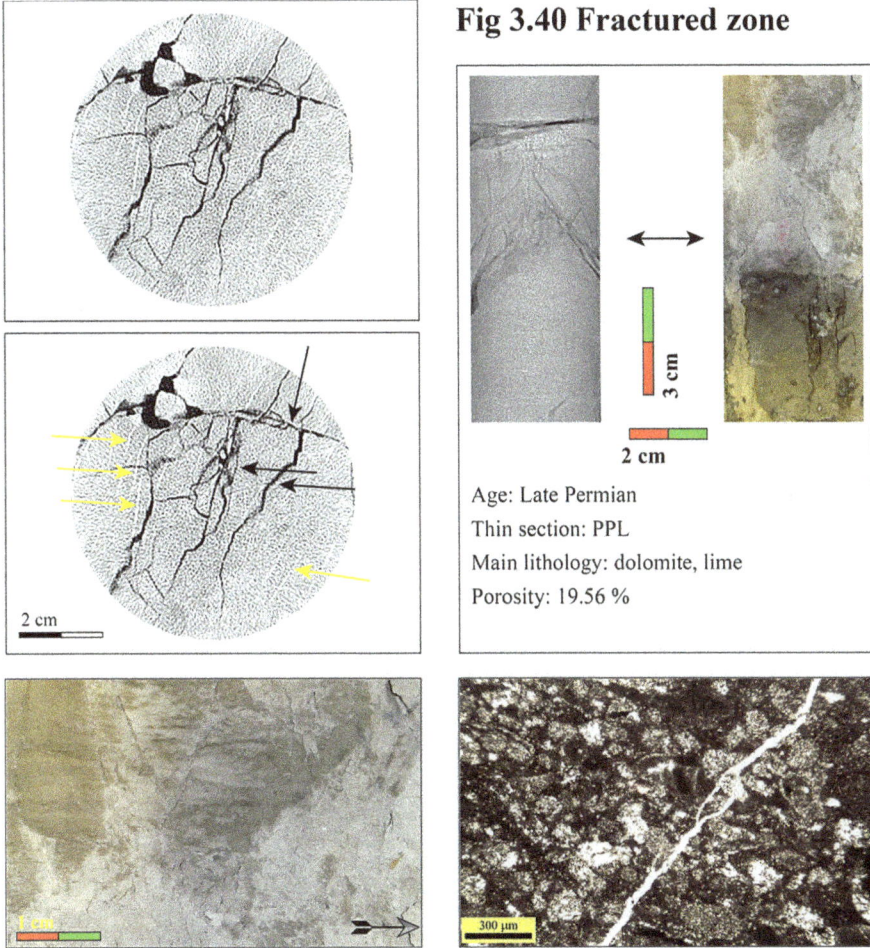

Fig 3.40 Fractured zone

Age: Late Permian

Thin section: PPL

Main lithology: dolomite, lime

Porosity: 19.56 %

FIGURE 3.40

Frequent filled (yellow arrows) and open fractures (black arrows) are present in this sample, which indicate an active tectonic zone. Anhydrite has been filled the fractures and so they are traced with white color in CT images. Probably, some fractures have been created by coring process, as they are not continuous at the core scale.

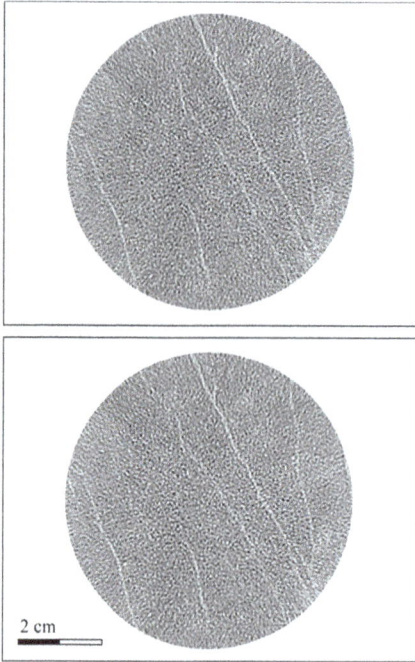

Fig 2.41 Fractured zone

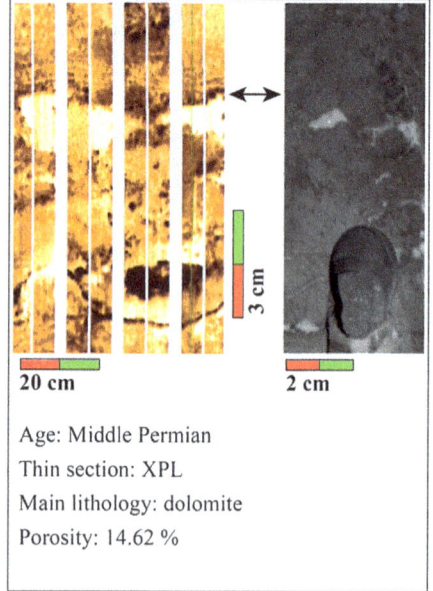

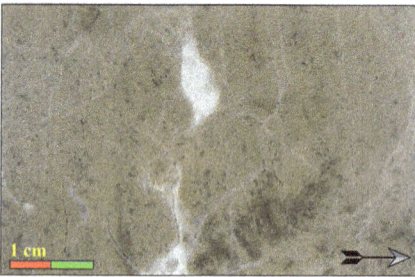

20 cm   2 cm

Age: Middle Permian
Thin section: XPL
Main lithology: dolomite
Porosity: 14.62 %

FIGURE 3.41

Fractures with various sizes and orientations are present in the sample. Based on the fracture orientation, some of them can be seen in CT images. Almost all fractures have been filled with anhydrite cements, which are white in CT slices. Large fractures are also visible in FMI log as sinusoidal lines.

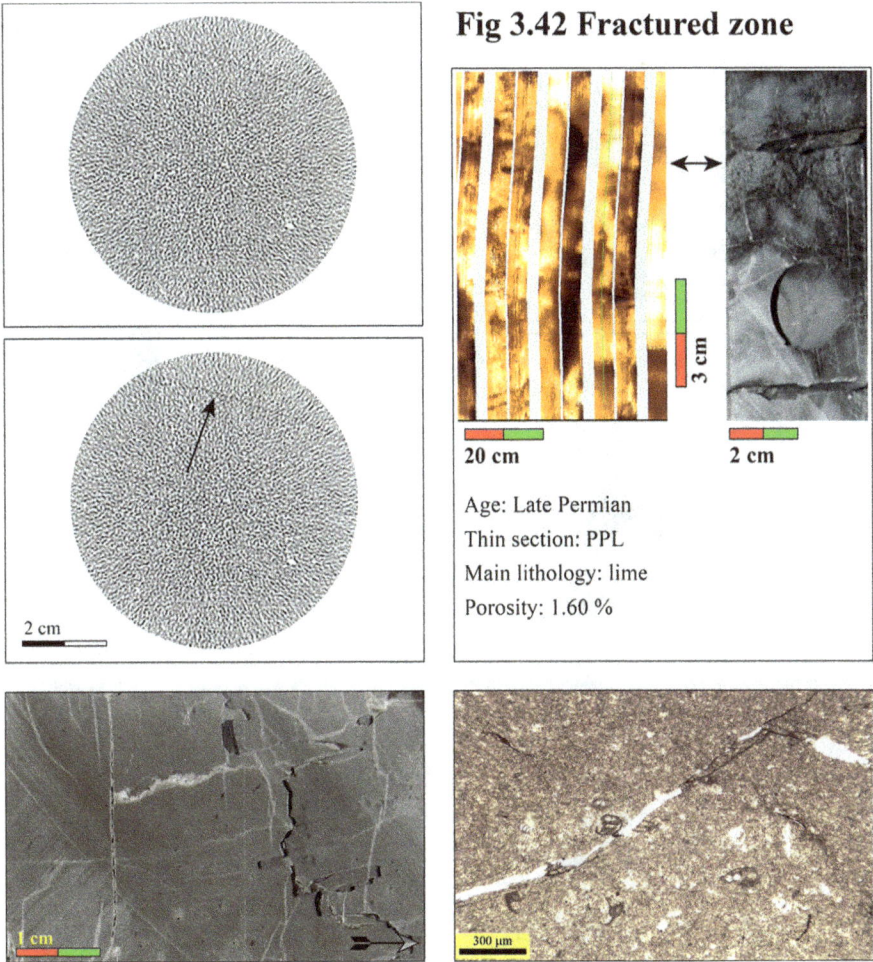

Fig 3.42 Fractured zone

Age: Late Permian
Thin section: PPL
Main lithology: lime
Porosity: 1.60 %

FIGURE 3.42
Many hairline fractures with small aperture sizes are visible in close-up and whole core photos, but they are not detectable in CT or FMI images. The CT slice just shows one open fracture (black arrow). Other fractures are not detectable because the CT and FMI resolutions are not enough for detecting these small features.

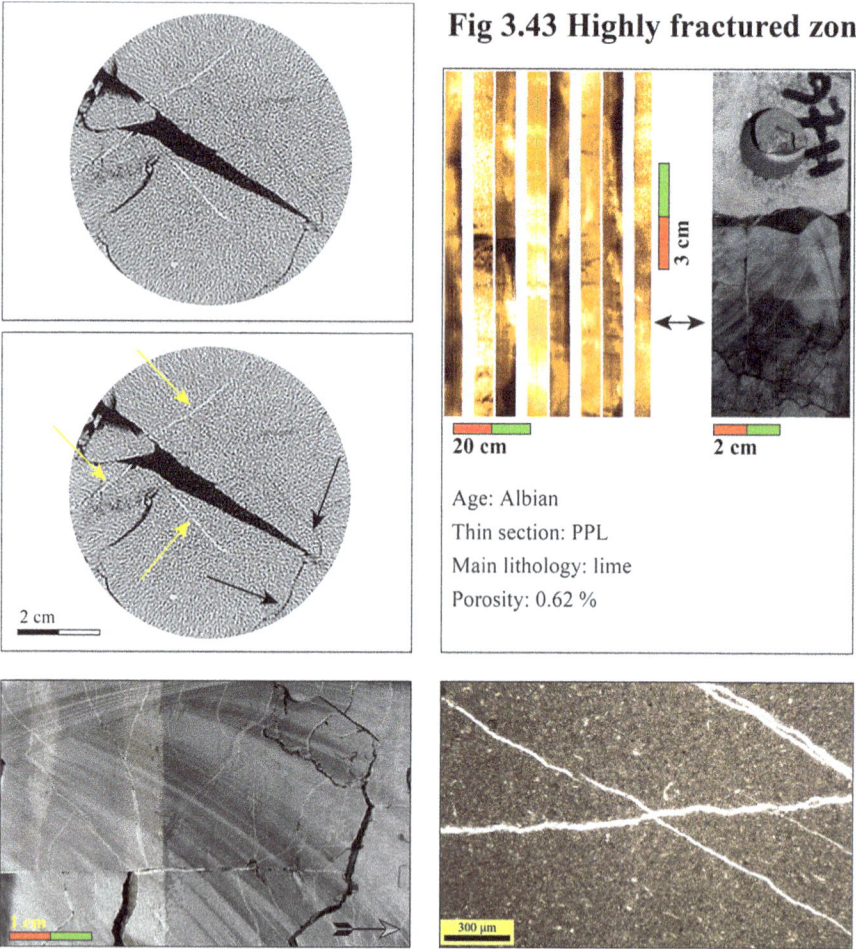

Fig 3.43 Highly fractured zone

Age: Albian
Thin section: PPL
Main lithology: lime
Porosity: 0.62 %

FIGURE 3.43
Both filled (yellow arrows) and open (black arrows) fractures are present in this sample. Filled fractures have been cemented by anhydrite in later diagenetic stages. Fractures are observable in whole core, close-up and thin section photos. The thick black part of the CT image has been created in coring process.

Plate 3.44 Natural hydraulic fracturing

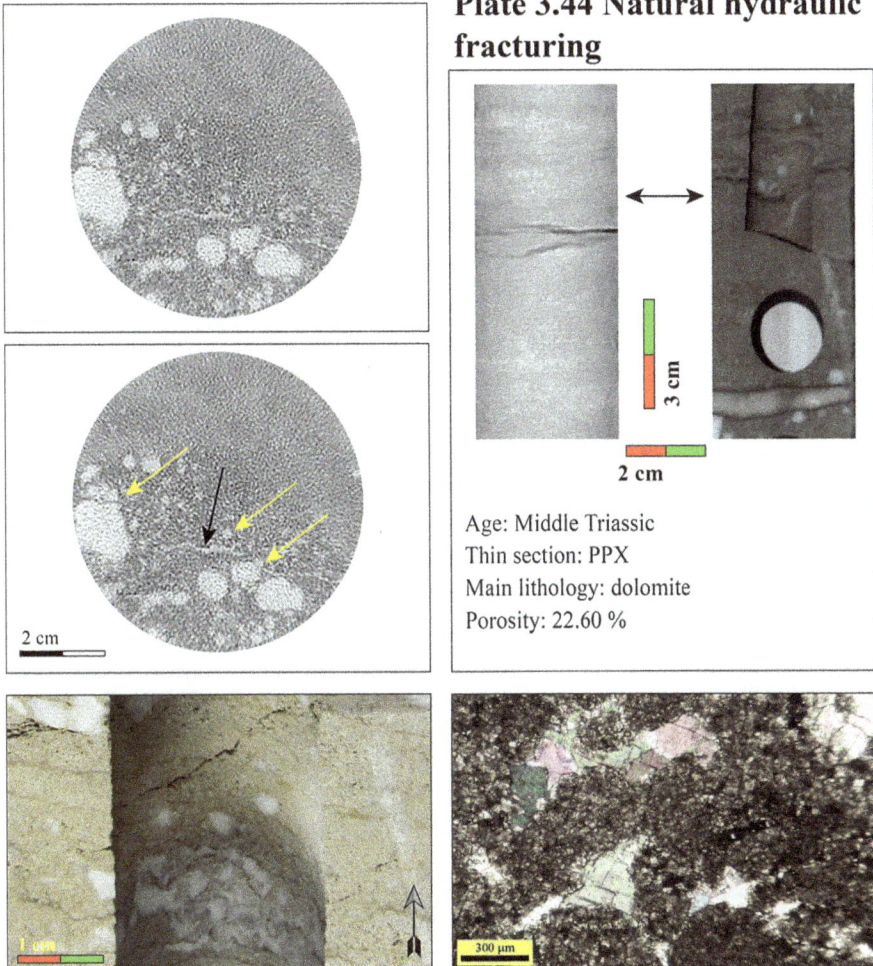

Age: Middle Triassic
Thin section: PPX
Main lithology: dolomite
Porosity: 22.60 %

FIGURE 3.44

Anhydrites are frequent evaporative minerals in carbonate reservoirs. They are formed from precursor gypsum by dewatering. The water increases the pore pressure and creates the fractures (black arrow). They are easily distinguishable because are always associated with large anhydrite nodules and cements (yellow arrows). These fractures are also visible on whole core and thin sections. They are routinely small and have no considerable effects on reservoir properties.

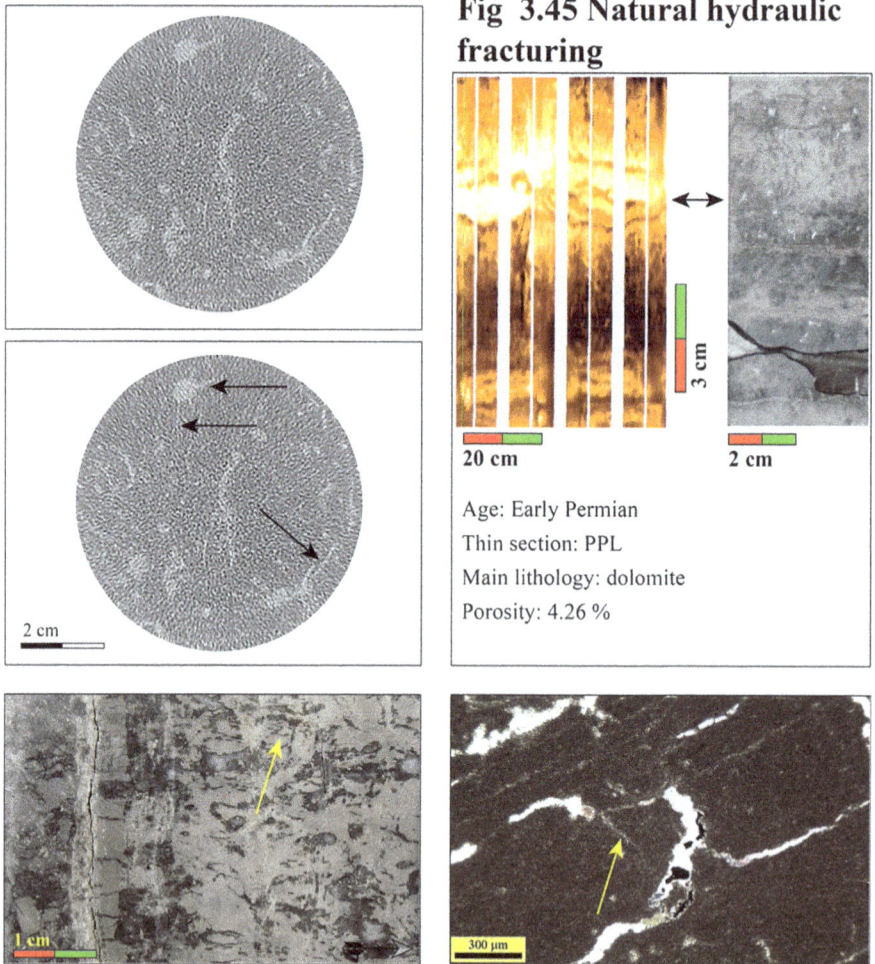

Fig 3.45 Natural hydraulic fracturing

20 cm   2 cm

Age: Early Permian
Thin section: PPL
Main lithology: dolomite
Porosity: 4.26 %

FIGURE 3.45
Natural hydrofractures are created by the pressure of a fluid in the reservoir rocks. There is no pathway for this fluid to exit and so pressure increases, which causes tiny fractures. In this figure, transformation of gypsum to anhydrite has produced water, which caused hydrofracturing (arrows). They are characterized by proximity to an anhydrite nodule and very small sizes.

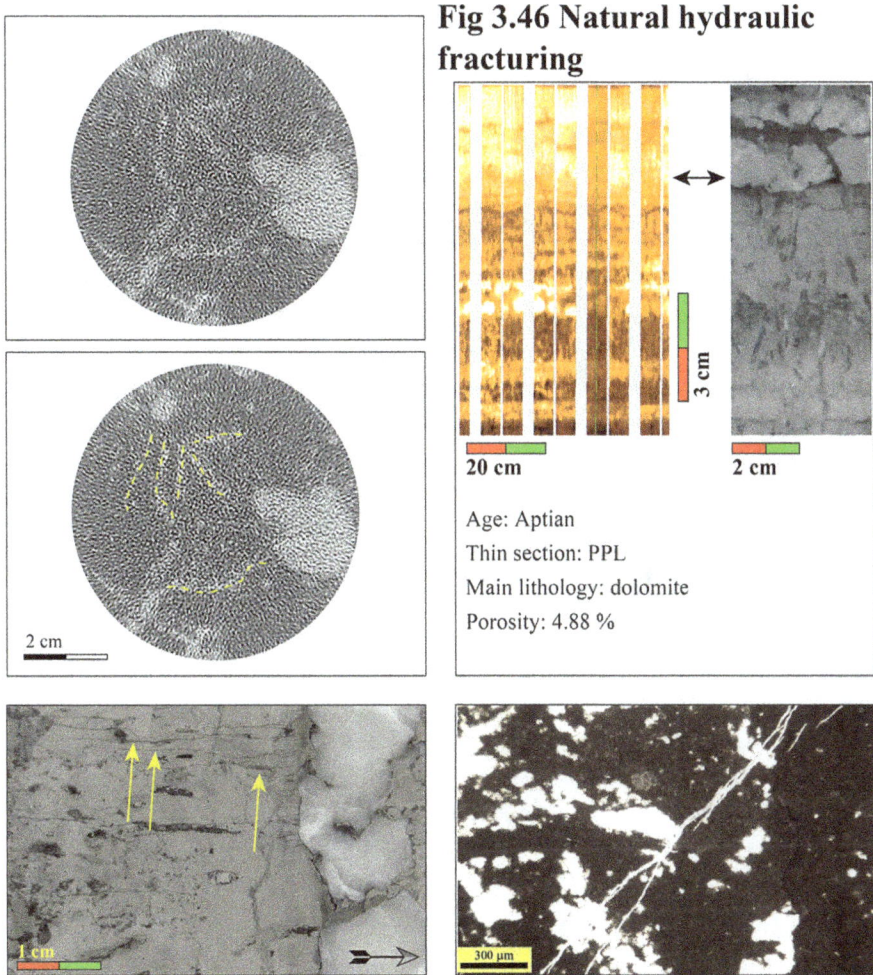

Fig 3.46 Natural hydraulic fracturing

Age: Aptian
Thin section: PPL
Main lithology: dolomite
Porosity: 4.88 %

FIGURE 3.46

Overpressure is caused by increasing fluid content of the rocks. The fluid is supplied by mineral transformations in many cases. In this sample, gypsum has been transformed to anhydrite, which resulted in overpressure in the rock. Therefore, natural hydraulic fractures have been formed (dashed line on CT image and arrows on close-up photos). The sample has high anhydrite content and so the fluids cannot escape from the rock.

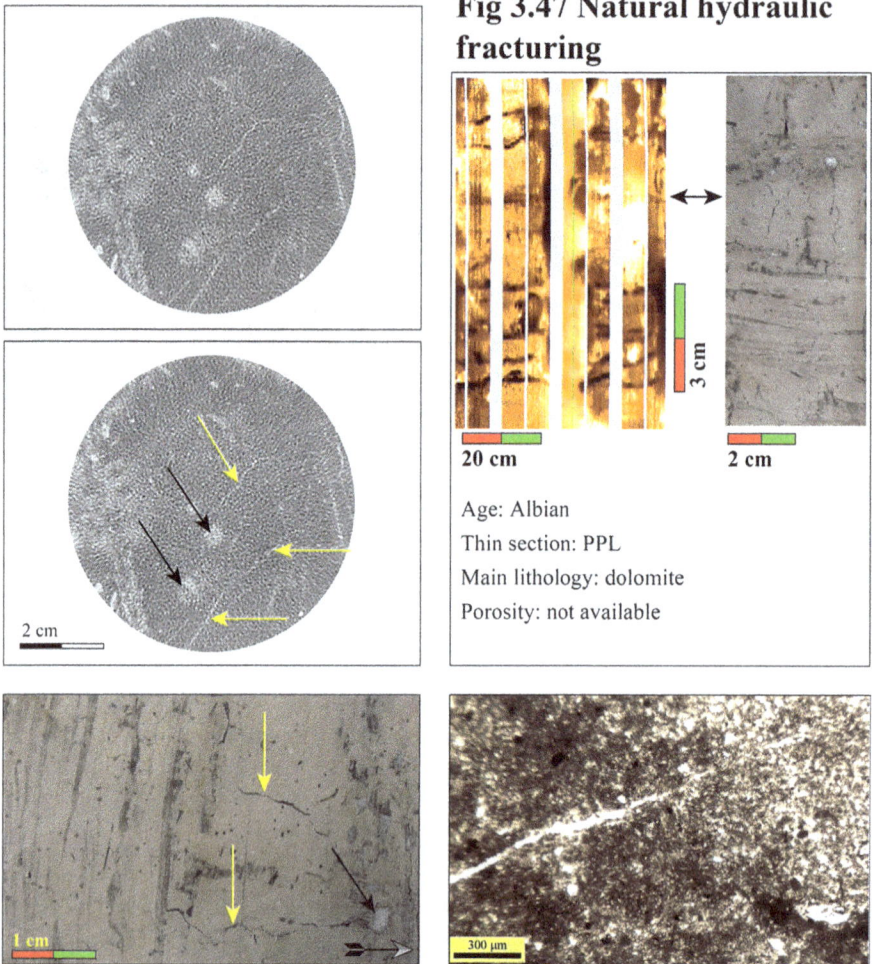

Fig 3.47 Natural hydraulic fracturing

20 cm   2 cm

Age: Albian

Thin section: PPL

Main lithology: dolomite

Porosity: not available

FIGURE 3.47

These small and thin fractures originate from overpressure, which, in turn, is produced by gypsum dewatering. Due to high frequency of anhydrite, the water cannot escape and so the rock is broken. These natural hydrofractures (yellow arrows) are always associated with anhydrite nodules (black arrows).

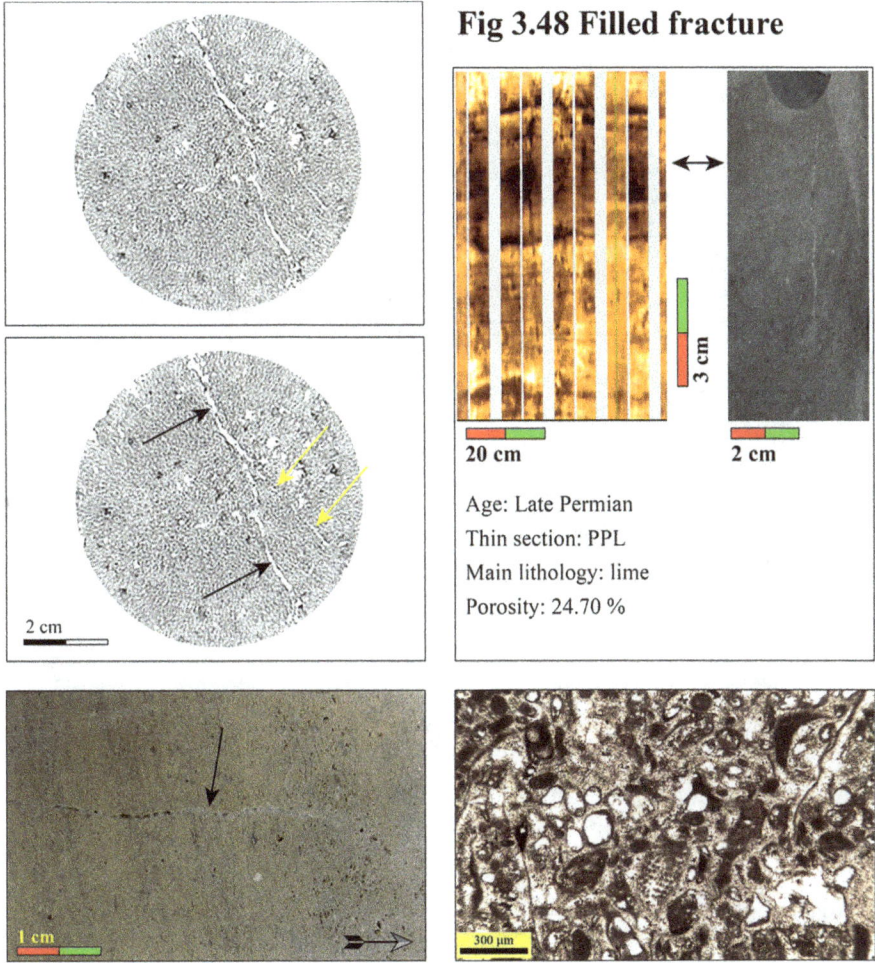

Fig 3.48 Filled fracture

Age: Late Permian
Thin section: PPL
Main lithology: lime
Porosity: 24.70 %

FIGURE 3.48

A vertical, semi-filled fracture is obvious in this CT image. The fracture-filling cement is anhydrite, and therefore it is marked by white color in this photo (black arrow). This large fracture is also visible in FMI, whole core and close-up (black arrow) photos. Another filled fracture is also traceable in CT image, which is not evident in other types of photos (yellow arrow).

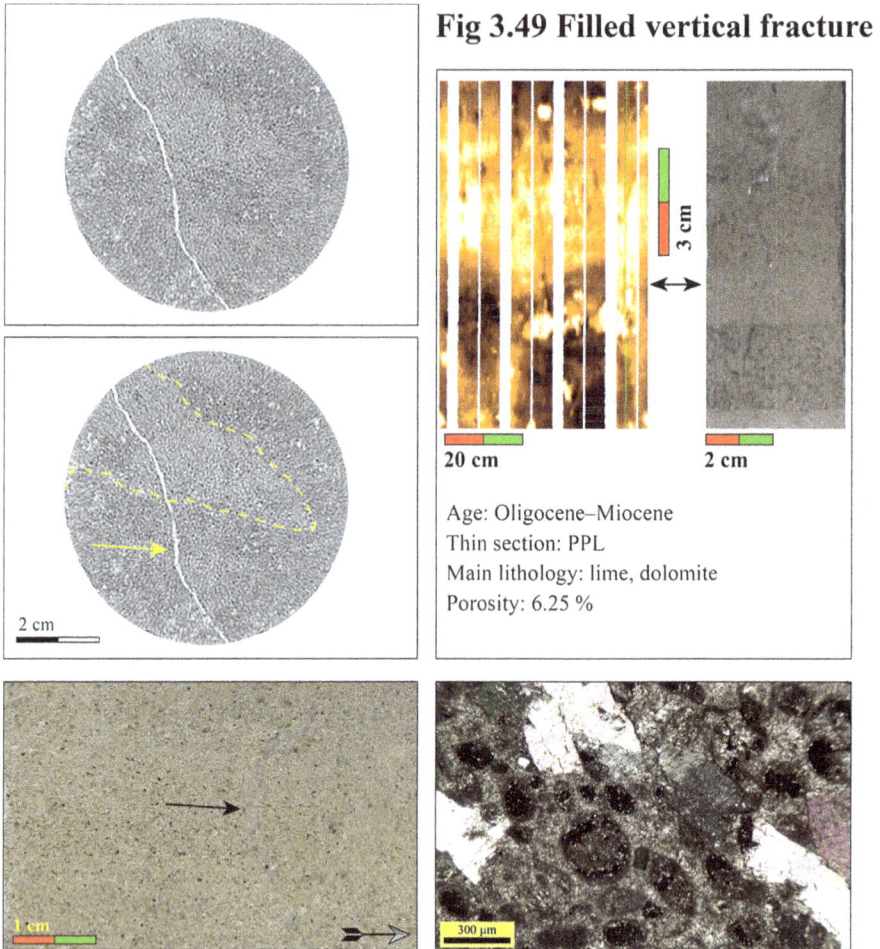

Fig 3.49 Filled vertical fracture

Age: Oligocene–Miocene
Thin section: PPL
Main lithology: lime, dolomite
Porosity: 6.25 %

FIGURE 3.49

Fractures are routinely filled with higher density minerals in comparison with rock matrix. In this CT image, a fracture is visible which has been filled with anhydrite during diagenesis (yellow arrow). This large fracture is also visible on whole core and close-up core photos, FMI image and thin section. Facies also changes at the slice location, which is evident with tongue-shaped change in the gray intensity of the image (dashed line). Small pore spaces can be recognized with dark spots in the transverse CT slice.

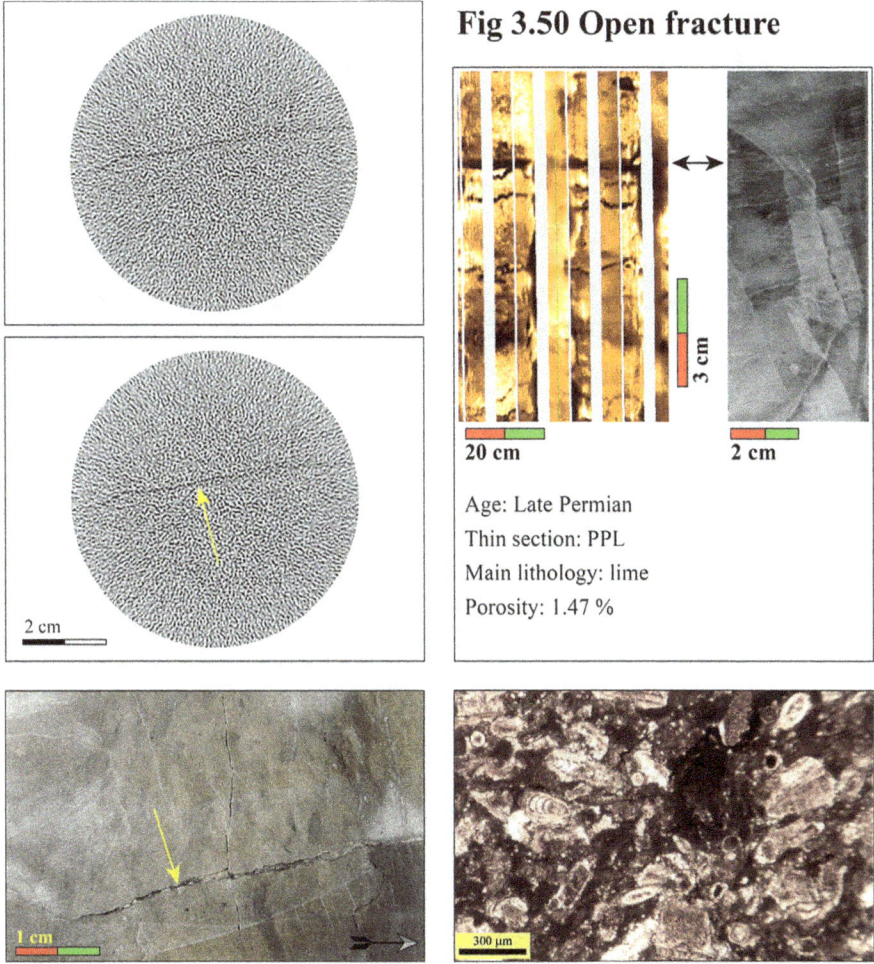

Fig 3.50 Open fracture

Age: Late Permian

Thin section: PPL

Main lithology: lime

Porosity: 1.47 %

FIGURE 3.50

A vertical open fracture is visible in this image. Open fractures are filled with fluids. Therefore, they are less dense compared to the matrix and are marked by dark gray shades in CT images (yellow arrow). Large vertical fractures are also detectable with dark color in FMI image (low resistivity). Aperture size can be determined in CT image.

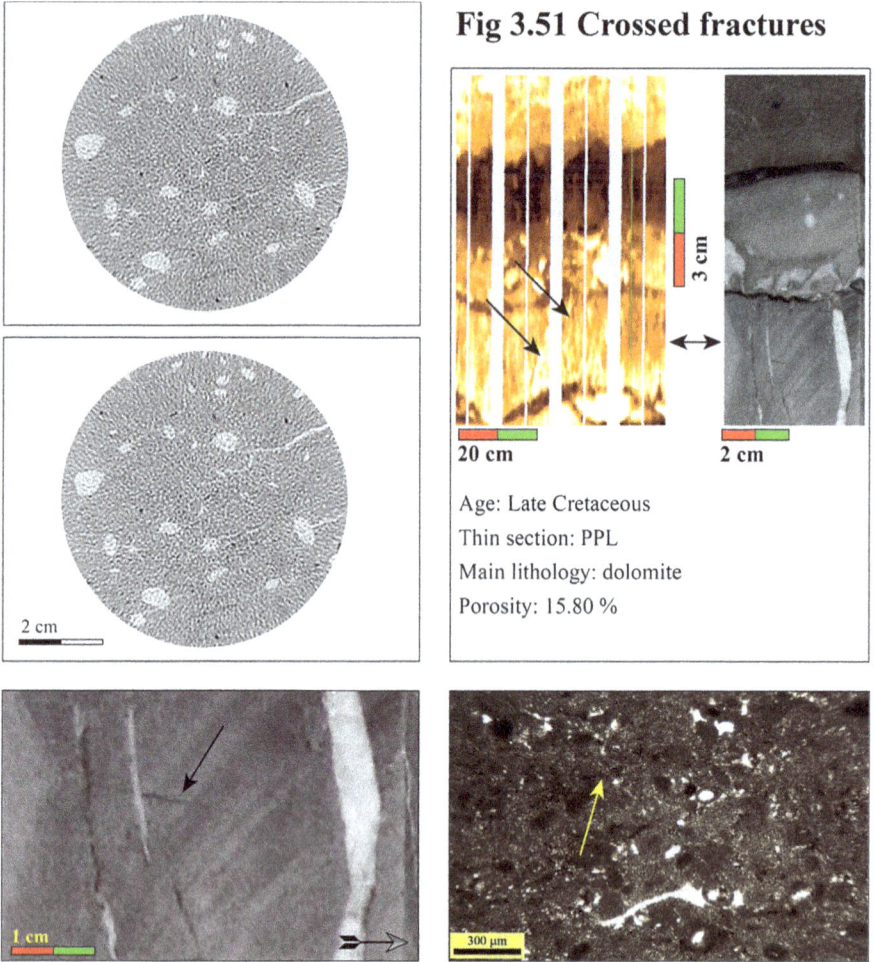

Fig 3.51 Crossed fractures

Age: Late Cretaceous
Thin section: PPL
Main lithology: dolomite
Porosity: 15.80 %

FIGURE 3.51
Both horizontal and vertical fractures have been developed in this sample (arrows). These fractures are visible in whole core and close-up photos but just vertical fractures can be detected on transversal CT image. Anhydrite nodules are seen as light round spots in CT slice. Vertical fractures are marked by black arrows on FMI image.

Fig 3.52 Crossed fractures

20 cm   2 cm

Age: Late Permian
Thin section: PPL
Main lithology: lime
Porosity: 0.87 %

FIGURE 3.52

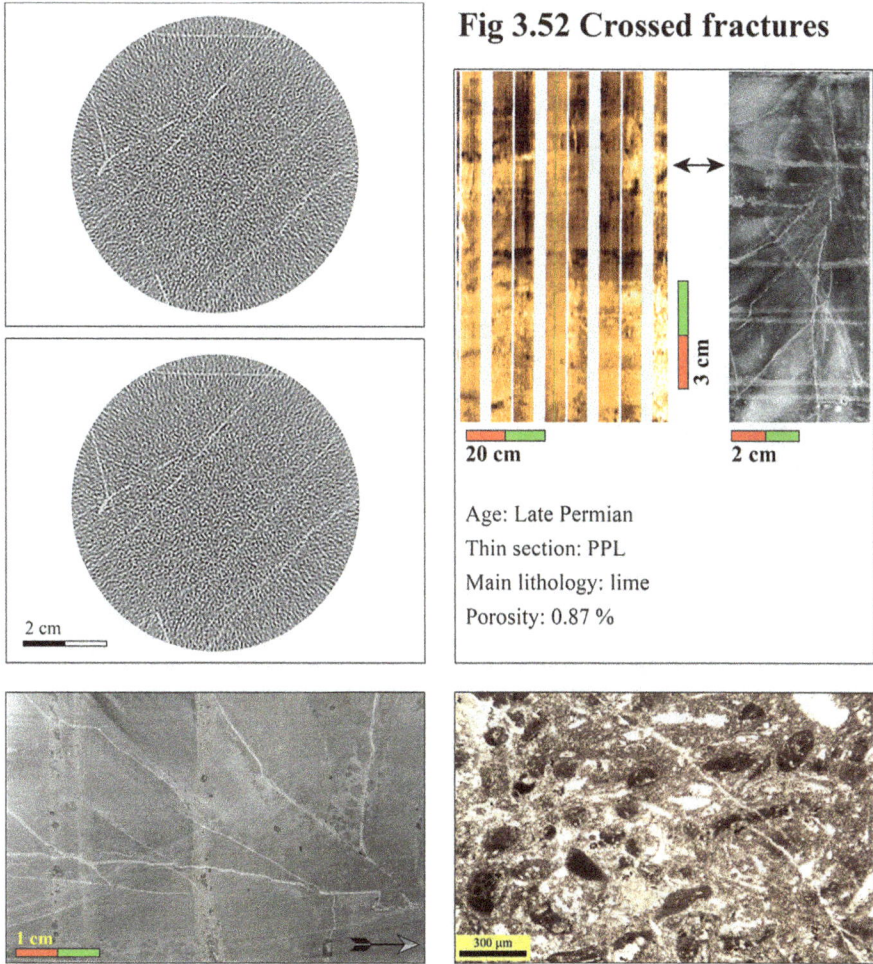

Fractures are seen in different directions and with various dips in this sample. All of them have been filled with anhydrite in late diagenetic stages. These fractures have been filled with anhydrite and so they are traced as white lines in CT slices. They are also visible in whole core and close-up photos. Large vertical fractures are also observable in FMI image.

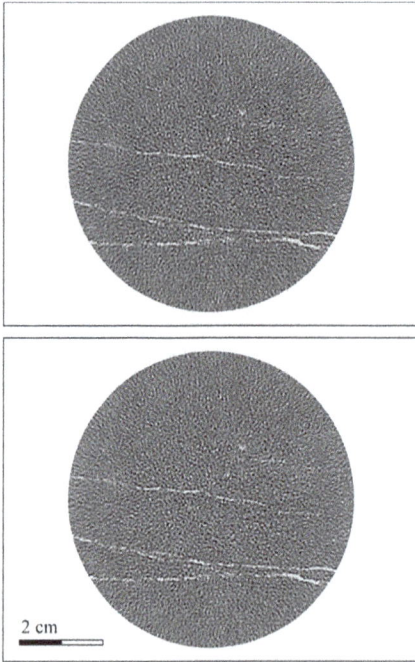

Fig 3.53 Crossed fractures

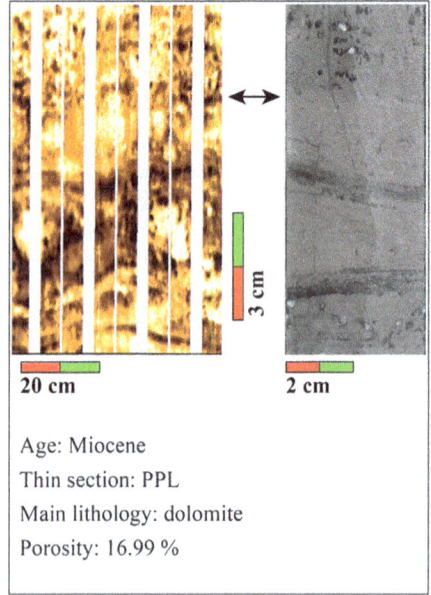

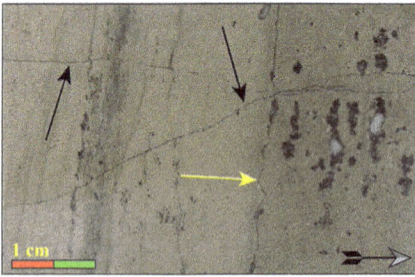

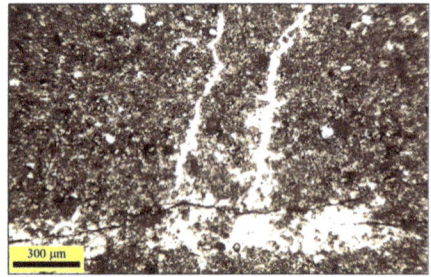

Age: Miocene
Thin section: PPL
Main lithology: dolomite
Porosity: 16.99 %

FIGURE 3.53
Vertical (black arrows) and horizontal (yellow arrow) fractures are obvious in whole core and close-up photos, but they cannot be differentiated in CT images. They have been filled by anhydrite in late diagenetic realm. Cross-laminations are also visible in whole core, FMI and close-up photos.

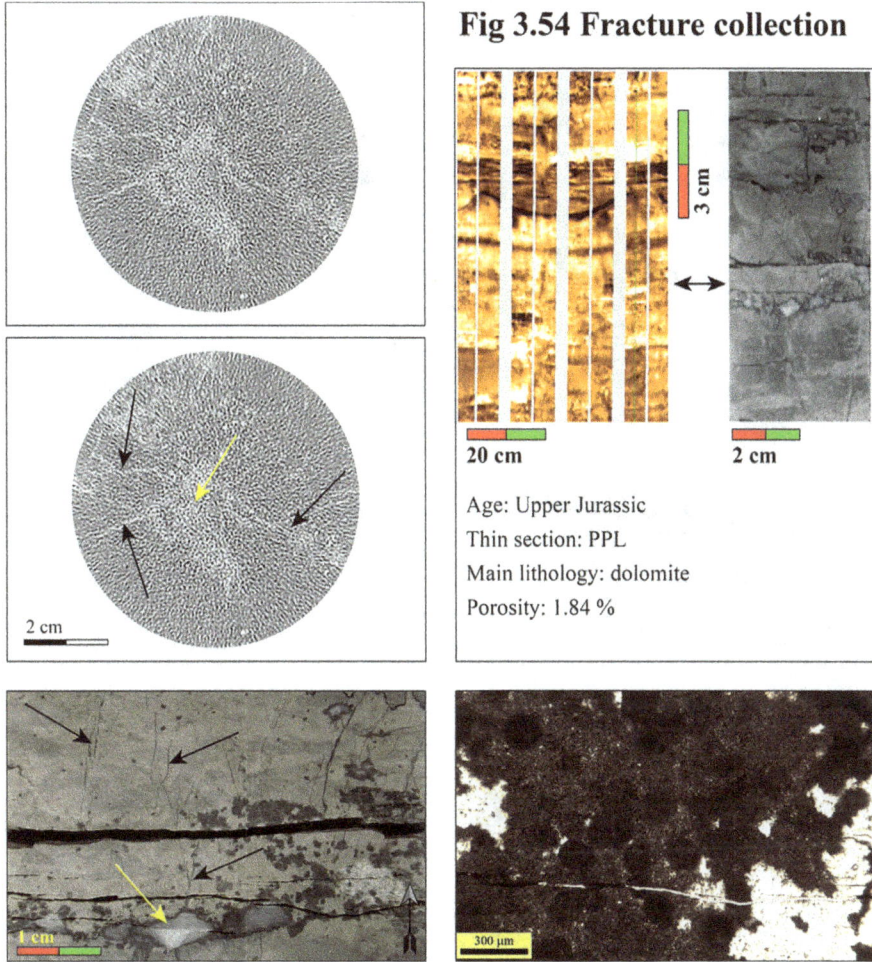

Fig 3.54 Fracture collection

Age: Upper Jurassic
Thin section: PPL
Main lithology: dolomite
Porosity: 1.84 %

20 cm

2 cm

3 cm

2 cm

FIGURE 3.54

A collection of natural hydrofractures (black arrows) is seen around an anhydrite nodule (yellow arrows). Water is expelled when gypsum is transformed to anhydrite in the diagenetic stages. This water causes overpressure which broke the rock and produced fractures. These fractures are filled with anhydrite in many cases.

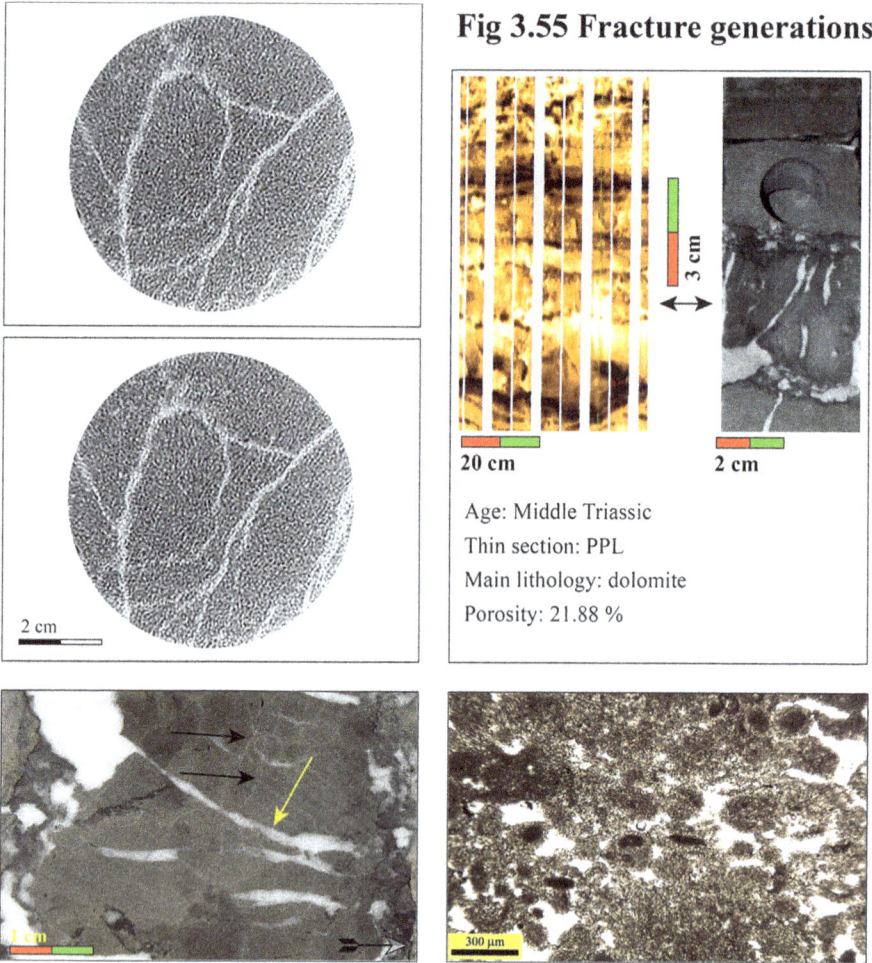

Fig 3.55 Fracture generations

Age: Middle Triassic
Thin section: PPL
Main lithology: dolomite
Porosity: 21.88 %

FIGURE 3.55

A highly dolomitized sample with various fracture generations (arrows) can be seen in transversal CT image. The presence of various fracture types with different directions is obvious on close-up photo of the core. Some of these fractures are also recognizable on whole core and FMI log. Their small aperture size makes it difficult to see some generations on FMI results. As thin sections cover a very small part of the rock, it is not always possible to trace them on thin sections.

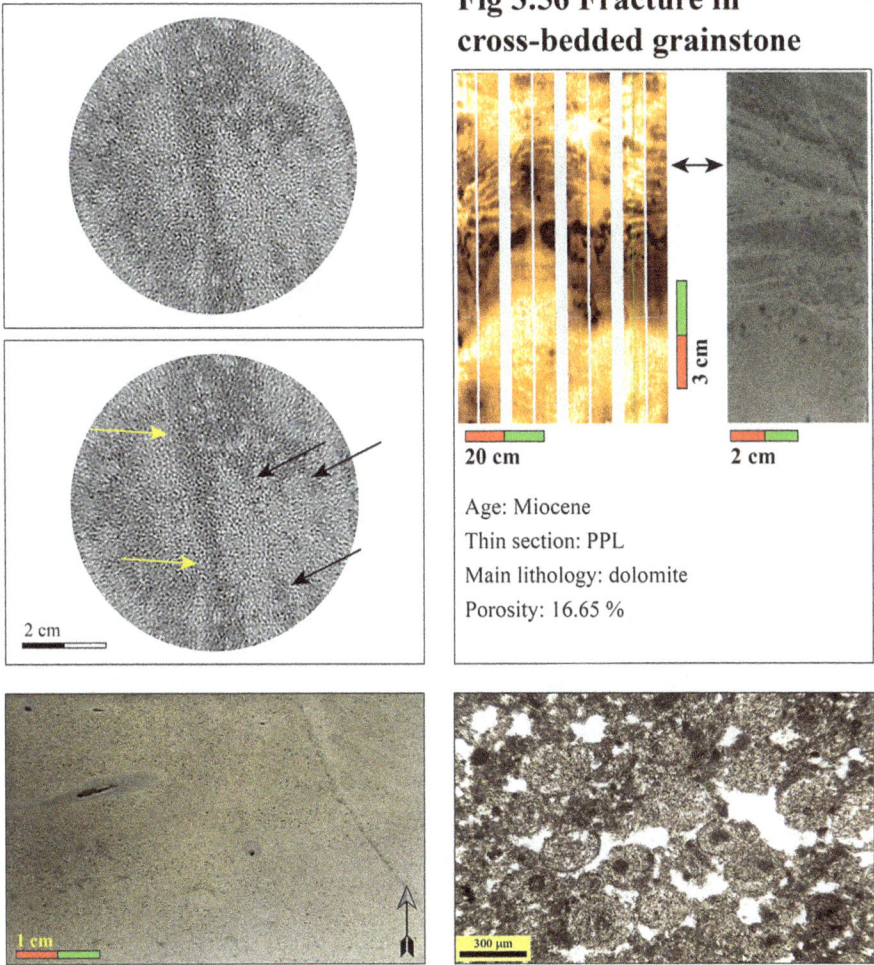

Fig 3.56 Fracture in cross-bedded grainstone

Age: Miocene
Thin section: PPL
Main lithology: dolomite
Porosity: 16.65 %

FIGURE 3.56
Open fractures are marked as dark lines while filled fractures are depicted as light gray or white lines in CT images. The alternation of dark and light gray colors in cross-bedded sample depends on the density of adjacent layers, which, in turn, changes with porosity and/or mineralogy. Both filled fractures (yellow arrow) and cross-beddings (black arrows) are obvious in this sample.

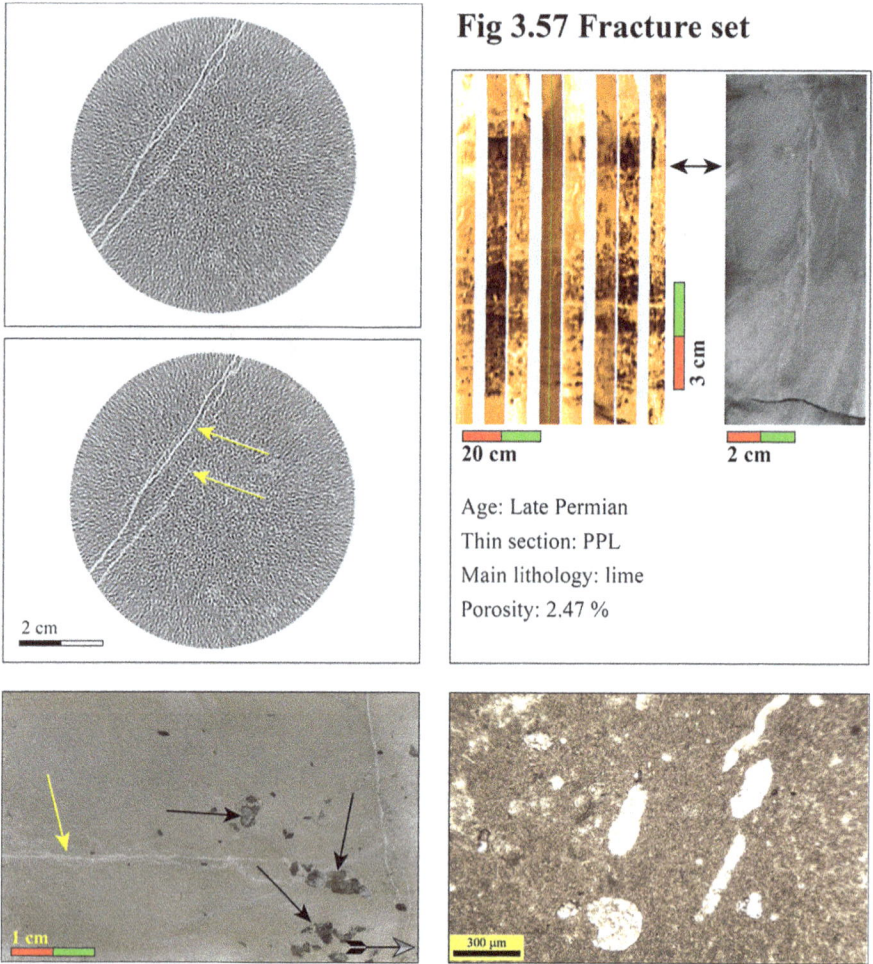

Fig 3.57 Fracture set

Age: Late Permian
Thin section: PPL
Main lithology: lime
Porosity: 2.47 %

FIGURE 3.57

A set of vertical fractures (yellow arrows) are marked by white color in CT scan image. Anhydrite pseudomorphs in the close-up core photo (black arrows) clearly show high anhydrite content of the sample. This mineral has filled the fractures after their formation. These fractures are not large enough to be detected by FMI log.

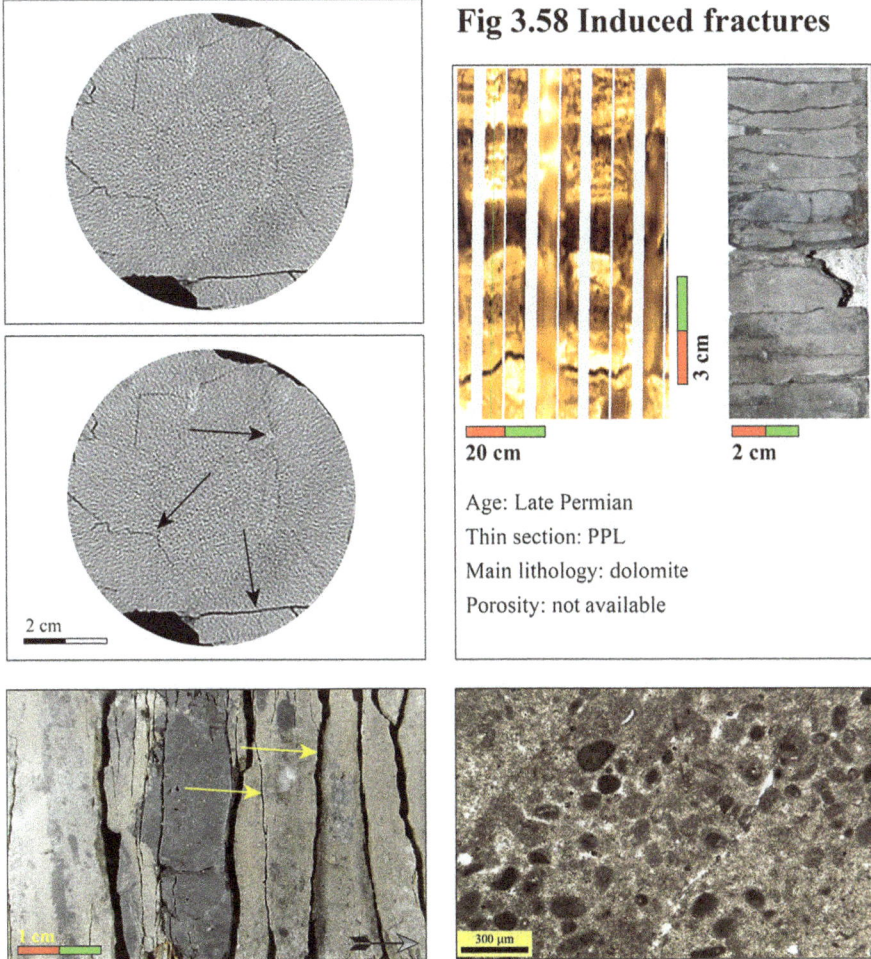

Fig 3.58 Induced fractures

Age: Late Permian
Thin section: PPL
Main lithology: dolomite
Porosity: not available

FIGURE 3.58
Induced fractures are created at the time of coring or after that in the layout process. These fractures are characterized by dark color and smooth surfaces in CT images (black arrows). Many of these fractures are perpendicular to the drilling direction (yellow arrows). These openings are created due to unloading at the boundary of the layers.

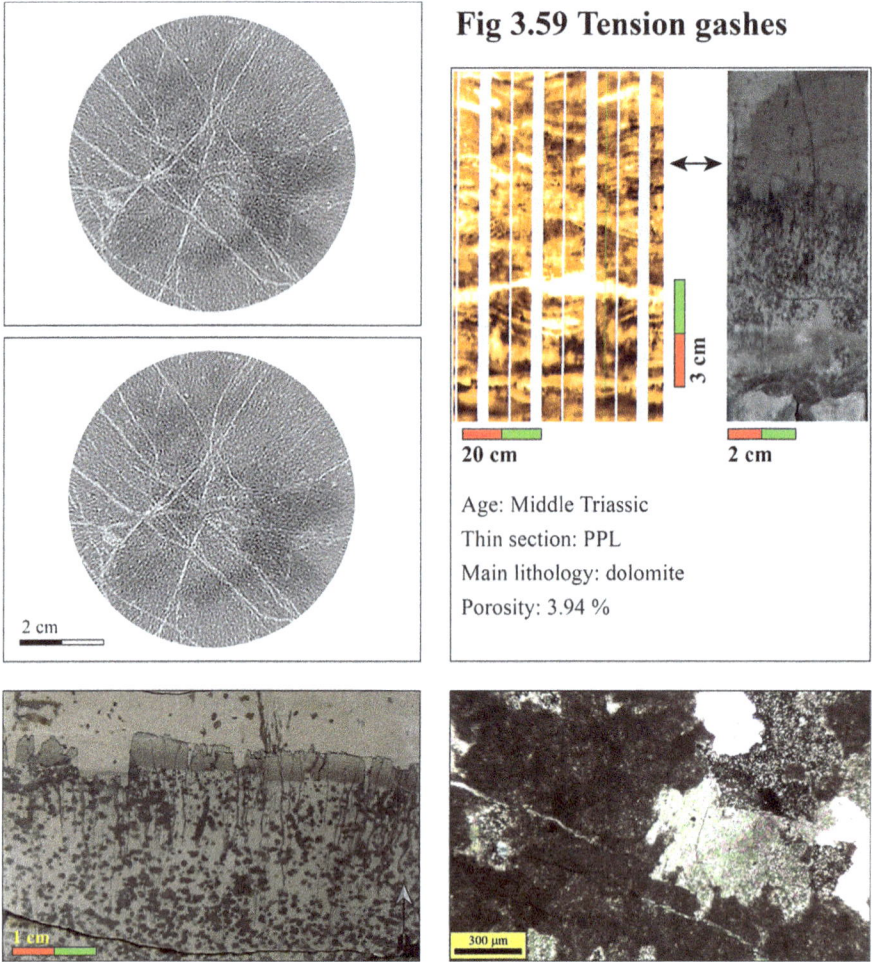

Fig 3.59 Tension gashes

Age: Middle Triassic
Thin section: PPL
Main lithology: dolomite
Porosity: 3.94 %

FIGURE 3.59

Tension gashes are formed perpendicular to the direction of maximum stress, which is vertical in many reservoirs. Many of them have been formed in this sample. Almost all these fractures have been filled by anhydrite in late diagenetic stages. Therefore, they are clearly recognizable with white color in CT image. Anhydrite pseudomorph and a compaction zone are also evident.

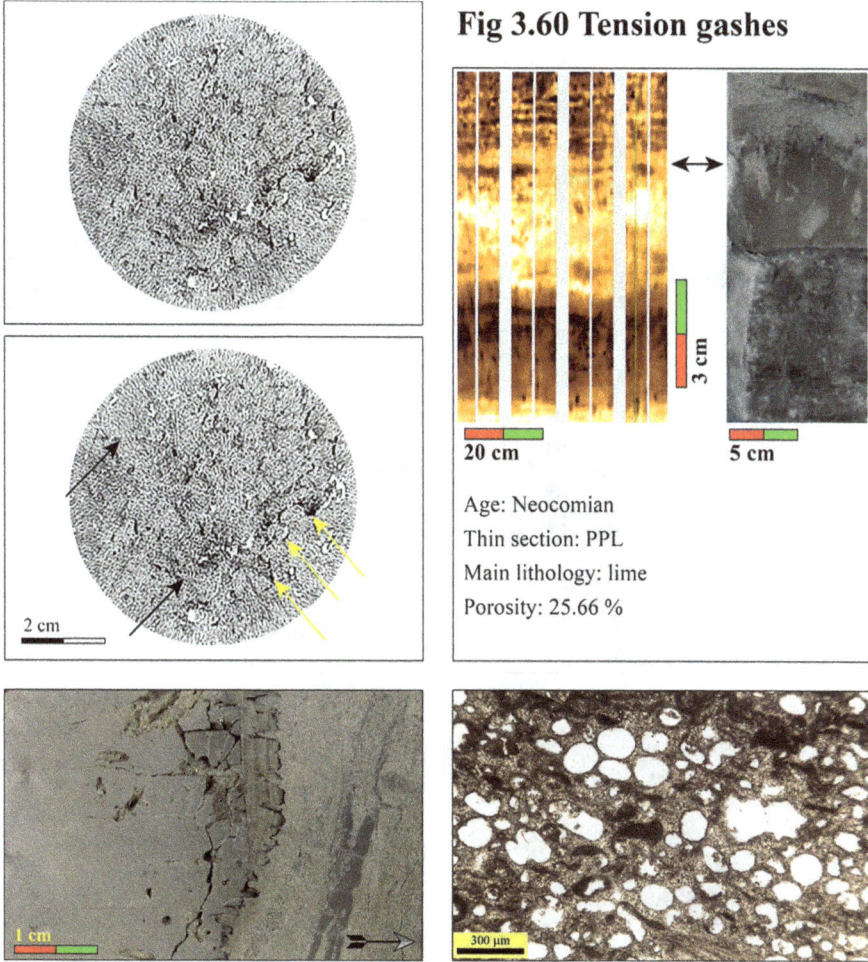

Fig 3.60 Tension gashes

Age: Neocomian
Thin section: PPL
Main lithology: lime
Porosity: 25.66 %

FIGURE 3.60

A low-amplitude, high-frequency stylolite is seen in this picture. Tension gashes are also present. Therefore, many smooth (black arrows) and jagged surfaces (yellow arrows) are traced as dark lines in CT image. The sample has high moldic porosity which is marked by black circles in CT image. Thin veins of anhydrite can be seen as white color areas in the image.

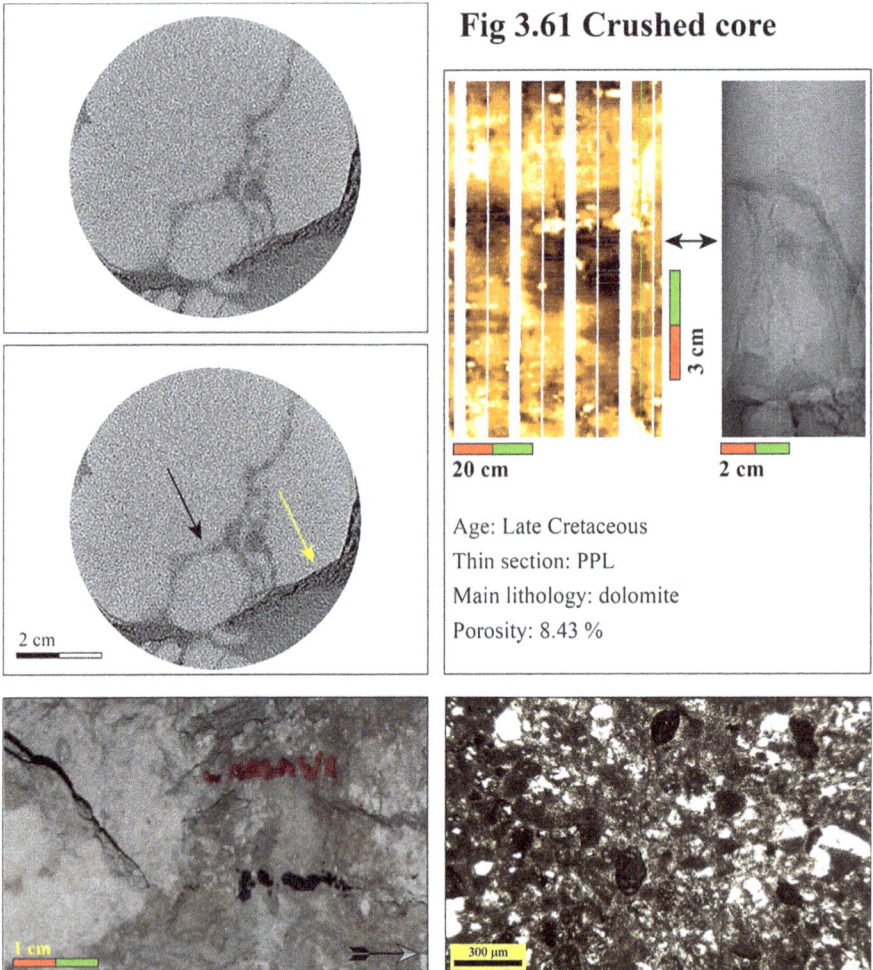

Fig 3.61 Crushed core

Age: Late Cretaceous
Thin section: PPL
Main lithology: dolomite
Porosity: 8.43 %

FIGURE 3.61

Cores are sometimes crushed at the time of coring. Usually, it happens in high porosity or fractured zones. This is very important for core layout because different treatments are necessary in this stage of core analysis. Low-density mud and mud filtrate invade the pore spaces or fractures. Therefore, they are seen as large areas with dark color along with crushed texture.

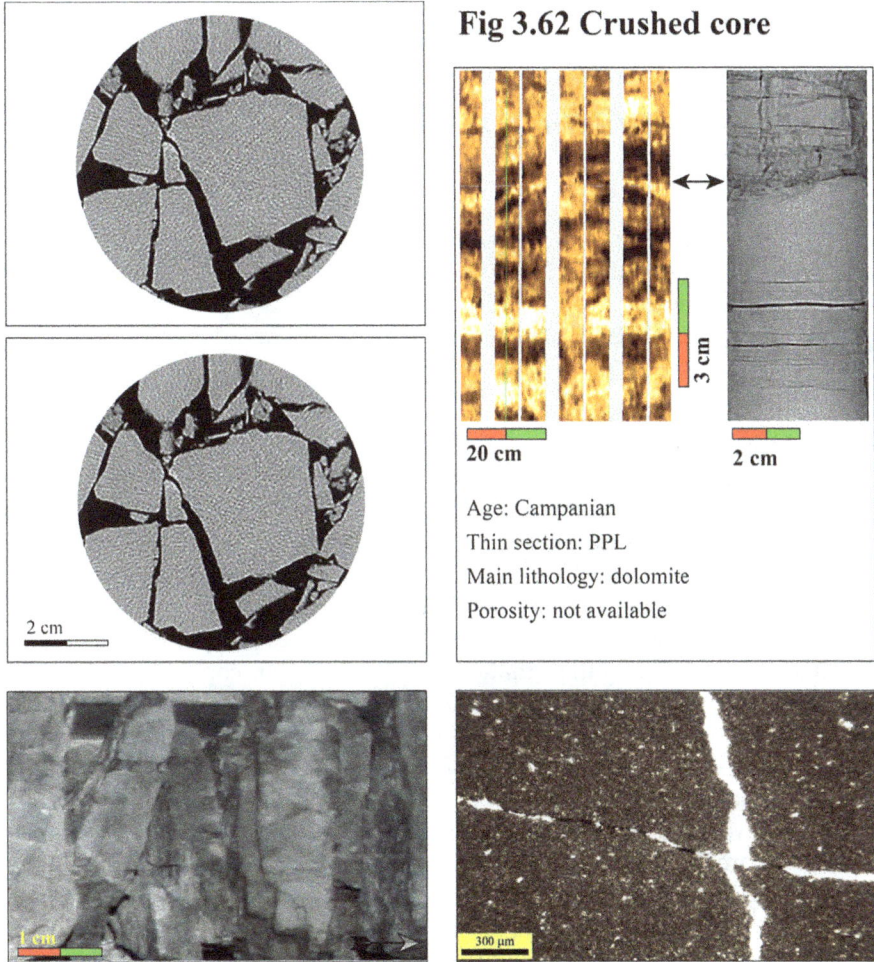

Fig 3.62 Crushed core

Age: Campanian

Thin section: PPL

Main lithology: dolomite

Porosity: not available

FIGURE 3.62

Many cores are crushed during coring procedure. This is not a natural process but natural fractures or high amounts of porosity intensify this process. Result is obviously seen on both transversal and longitudinal core CT scan images. Sample preparation is not possible from these intervals and so petrophysical data is not available.

4

Image Enhancement

4.1 Introduction

CT scan images are two-dimensional arrays of picture elements or pixels. In a color image, each pixel has tree values for red, green and blue and combination of these colors form the final image. In contrast, each pixel has just one value in CT images. This value changes between 0 and 255. In digital images, colors are defined by the intensity of light and so, 0 indicates black while 255 shows white color in these pictures. There is a range of gray intensities between these two end members. These values are recorded based on the density of the examined material. The maximum contrast can be seen between various material phases (solid, liquid and gas). Therefore, empty spaces (filled with air) are marked with dark color in these images. In some tests, the sample is saturated with a fluid. These fluids are also darker than the surrounding solid rock matrix in CT scan photos.

Frequency of gray intensity of pixels in a CT image shows the heterogeneity of the sample. It also indicates special features in some cases. Therefore, the first figure of this chapter shows a CT image and its histogram of pixel values (Figure 4.1). Normal distribution of values indicates low heterogeneity while bimodal or multimodal distributions show high frequency of special features in the photo. Histograms are also useful for distinguishing lithology. Dense minerals in carbonate reservoirs, such as dolomite or anhydrite, are light in color in CT images. Therefore, higher values are seen in the histogram of gray values distribution of these images. In contrast, CT images of samples with lower density materials, such as calcite, clays or porosity, are darker and so their histograms show higher frequency of lower values. These distributions also depend on brightness and contrast of the images. In CT scan images of the reservoir cores, the main purpose is distinguishing geological and petrophysical features. Therefore, the brightness and contrast are changed for better view. Changes in these two parameters also affect other properties of the image, including the distribution of gray intensities. Therefore, one image has been selected to show the effect of these changes on CT images (Figures 4.2 and 4.3).

DOI: 10.1201/9781003405412-4

Selecting an intensity threshold is a routine process in CT image analysis. Such selection is used to extract geological and petrophysical features of these images. Examples are porosity and mineralogy. Selecting a threshold is also used in binarization of the images. Therefore, porosity and anhydrite content of a CT photo have been extracted in the following pages (Figure 4.4). Pixels with almost equal intensity can also be selected using Magic Wand tool (Figure 4.5). This tool is available in many image analysis software. Tolerance in the values is determined by the user. Higher amounts lead to the larger areas and vice versa. This is more useful for selecting patchy objects with variable gray intensities, as this is performed by the user. Distinguishing the difference between gray pixels is not easy. So, the gray spectrum of CT photos can be replaced with other colors. The result is a CT photo with pseudo colors (Figure 4.6). The range can also be replaced by just two colors namely black and white. This process is called binarization. Depending on the selected threshold, these two colors show various features. Porosity and mineralogy are two common parameters (Figure 4.7). Instead of just two colors, all range of gray-level data can be transformed to a number of groups. This process is called clustering and can be performed by several methods. The result of K-means clustering is presented in this chapter (Figure 4.8). In this method, the center of the cluster is used for distance calculations and pixels with nearest distances are grouped together.

Sometimes bitmap images are not appropriate for image analysis. Therefore, they are converted to vectors. A threshold is selected and pixels with lower values are traced. Different algorithms are available and various thresholds can be selected, which determine the final result of the process. Traced lines are continuous and so some petrophysical properties (such as permeability) can be calculated more easily. Vectorizing with different algorithms and thresholds are presented in this chapter (Figures 4.9–4.12). It is worth mentioning that image enhancement is not limited to these techniques. Many filters are used to enhance or modify images before further analysis. They include, but are not limited to, median, Gaussian blur, Laplacian, neighborhood average, sharpen, blur and smooth (Figure 4.13). The first two are presented in this chapter. Median filter replaces pixel values of a CT image by the median value of neighboring pixels (Figure 4.14). Number of these pixels is determined by the user and so the results will be different. This filter can be used to remove image noises. Gaussian blur removes details and noises of an image using a Gaussian distribution function (Figure 4.15). In addition to filters, mathematical calculations are used to enhance the CT images. For example, an image can be added to itself (Figure 4.16). All pixels move toward the lighter values in this process, except 0 values. Therefore, there will be more contrast between pore spaces and other parts of the images (Figure 4.17). Edge detection is also useful for distinguishing allochems and porosities. The boundaries of the objects are marked in this process (Figure 4.18). The last figure of this chapter shows the effect of metal objects on CT scan of the core samples (Figure 4.19).

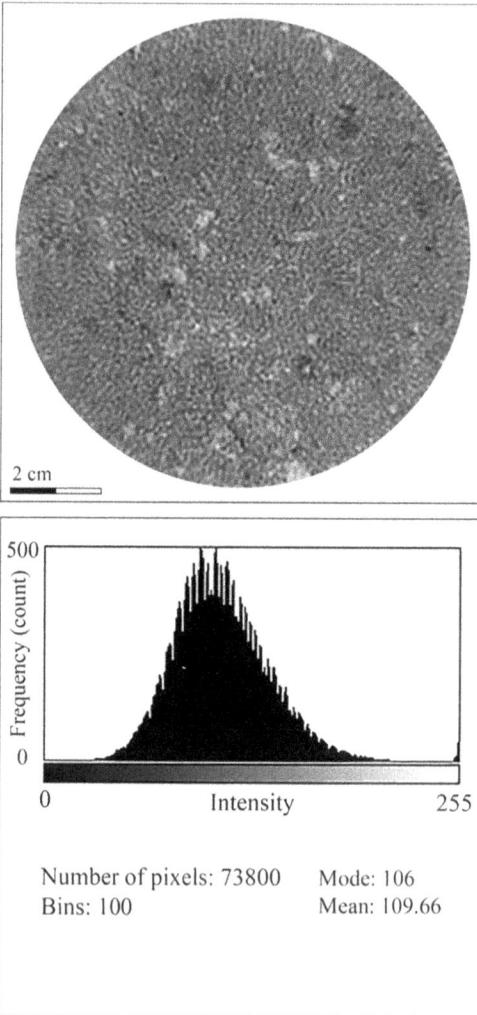

Fig 4.1 Gray-level data

Number of pixels: 73800   Mode: 106
Bins: 100   Mean: 109.66

Age: Neocomian
Main lithology: dolomite
Porosity: 22.60 %

FIGURE 4.1

Most of the image analysis techniques calculate the gray intensity of pixels in a gray-scale CT image. Therefore, it is very important to know the frequency distribution of pixel values in a CT image. The upper figure shows a CT image and the lower image is the plot of gray frequency in this image. Both pore spaces (dark parts) and anhydrite nodules (light parts) are present and so the sample has a normal distribution of various intensities of gray color. As can be seen, the frequencies of very high (near to 255) or very low (near to 0) gray intensities are low. Mode is 106 and mean is 109.66 (gray intensity), which indicate higher frequency of light colors. This is due to the high anhydrite content of the sample. Also, the background lithology is dolomite, which has high density and so shows lighter colors.

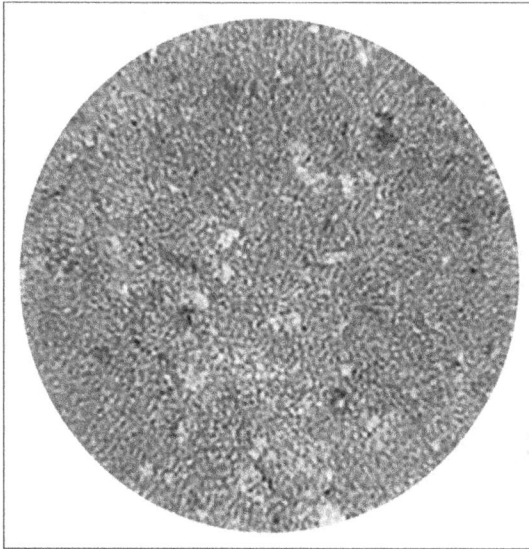

Fig 4.2 Brightness

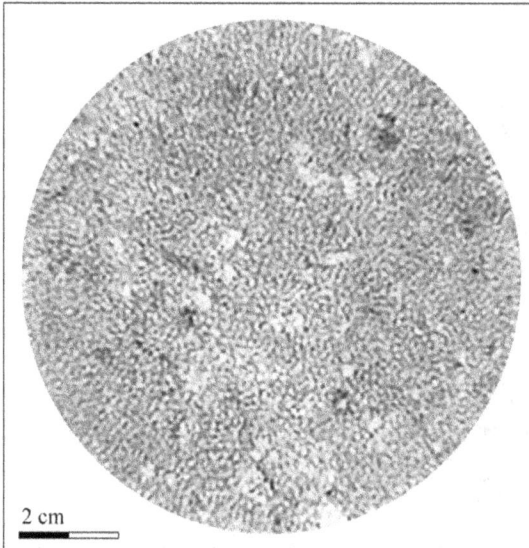

Age: Neocomian
Main lithology: dolomite
Porosity: 22.60 %

FIGURE 4.2
Brightness is darkness or lightness of a photo. It is more important in gray-scale images because it changes the intensity of the colors too. In this figure, the lower image is 50% lighter than the upper one. Both white and dark spots are still visible, but the overall lightness has been increased. Increasing the brightness of CT images increases the population of white or nearly white pixels. This affects the calculations which are based on the pixel colors in the slice. It is also used for better graphical view and finding the image features.

Fig 4.3 Contrast

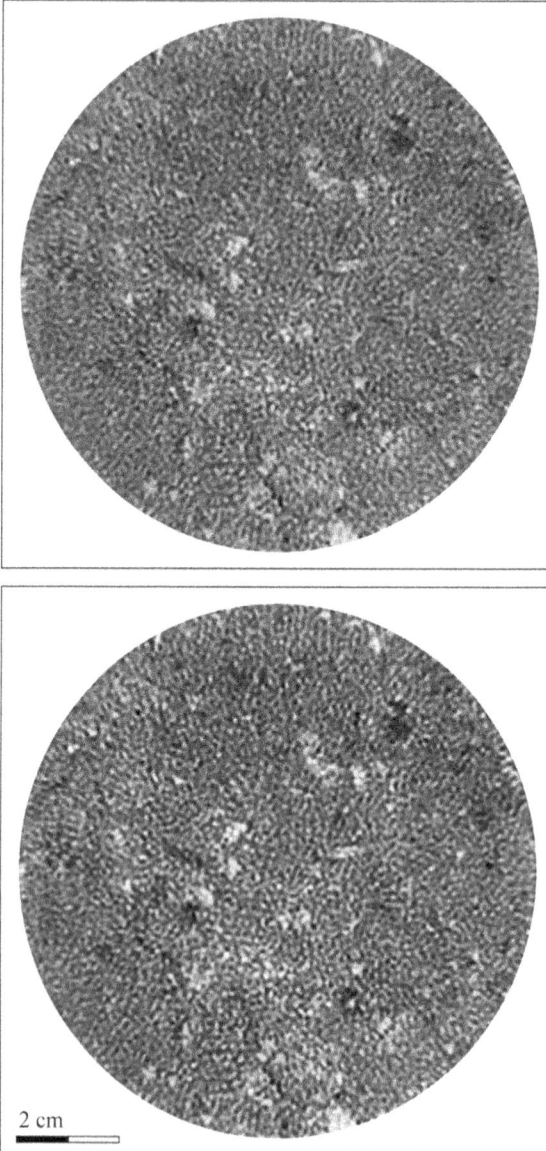

Age: Neocomian
Main lithology: dolomite
Porosity: 22.60 %

2 cm

FIGURE 4.3
In a CT image, contrast defines difference between gray color intensities of the pixels. Image contrast is 50% higher in the lower image. It can be seen that dark and light parts are still recognizable, but there is more difference between gray color intensities in the lower image. In fact, dark gray pixels became darker and light pixels became lighter. Therefore, the image has a better view in some cases. Increasing the contrast of an image makes it more pixelate and so it should be used with care.

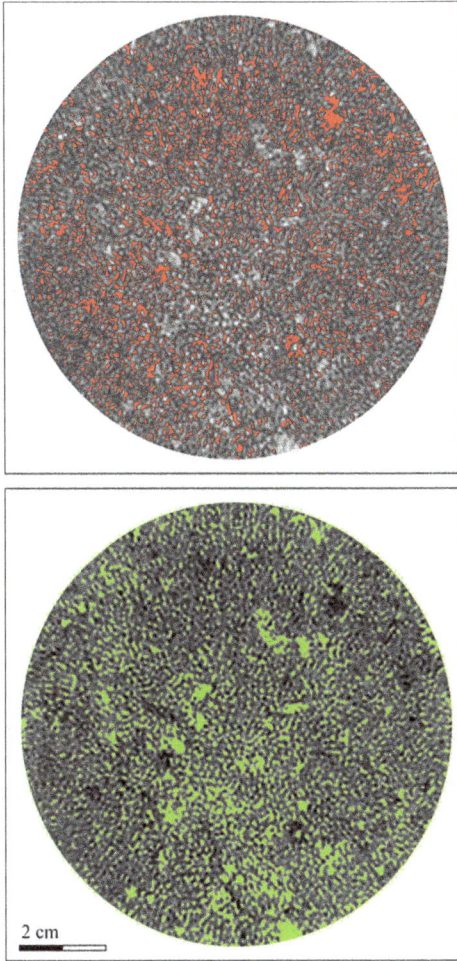

Fig 4.4 Intensity threshold

Age: Neocomian
Main lithology: dolomite
Porosity: 22.60 %

FIGURE 4.4
The threshold of color intensity (image lightness or darkness) helps in object extraction of a CT slice. The pixel value is defined by the intensity of light in CT scanning and so 0 intensity indicates black and 255 shows white color. Accordingly, gray pixel values between 0 and 80 have been highlighted in red in the upper photo, which comprises 19.13% of the whole image. It is close to the helium porosity of the plug sample, which is 22.60%. Anhydrite mineral has more density than surrounding dolomite. Here, 140–255 gray intensities have been used for calculating the anhydrite content in the sample (lower figure). Result is 22.12% of the whole image, which is close to the visual estimation of this mineral in relevant thin section (20%).

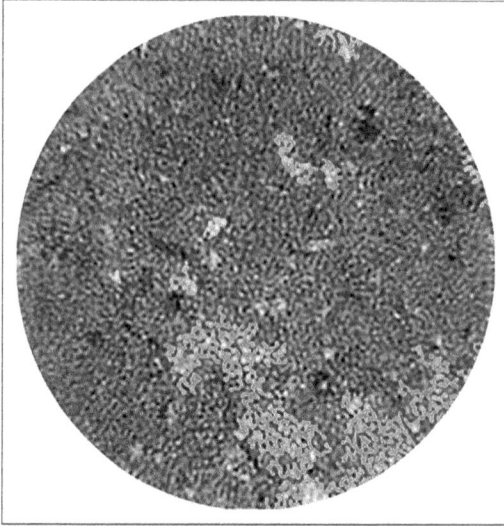

Fig 4.5 Magic Wand

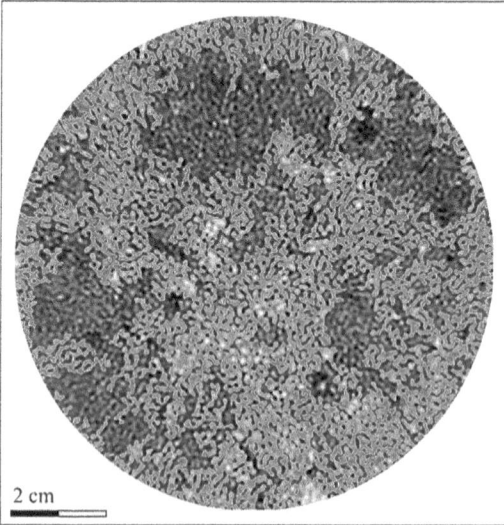

Age: Neocomian
Main lithology: dolomite
Porosity: 22.60 %

FIGURE 4.5

The Magic Wand tool selects objects with similar color. The selected area expands to the contiguous pixels with similar color. The threshold of similarity can be controlled by adjusting the tolerance values. Every single object should be selected separately, although some analyzing software can select the similar pixels automatically. Here, some porosities (black colors) have been selected by this tool. The tolerance of 20 (gray intensity) has been selected for the upper image, while it is 30 in the lower photo. The difference in the selected areas is obvious.

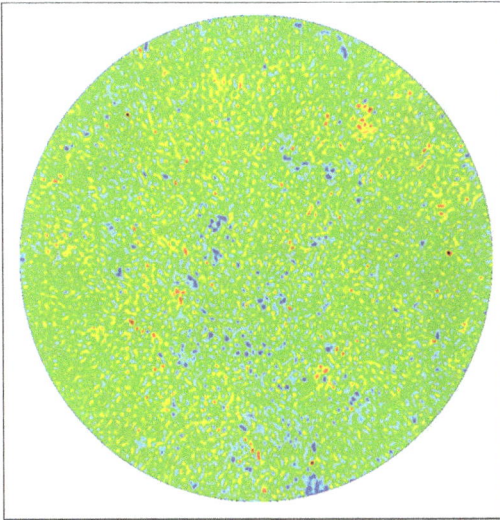

Fig 4.6 Pseudocolor

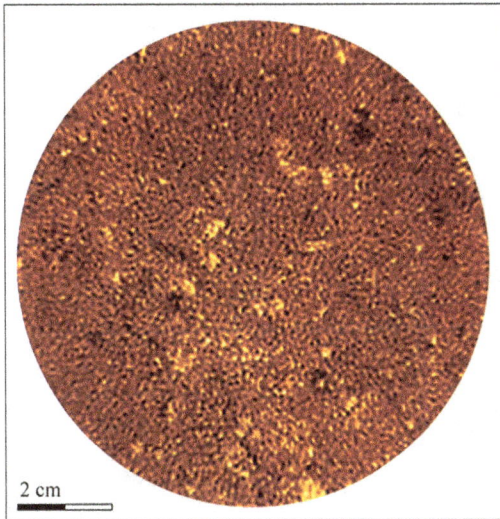

Age: Neocomian
Main lithology: dolomite
Porosity: 22.60 %

2 cm

FIGURE 4.6
Colors could be replaced by other colors in every image, but this is more important in gray-scale CT scan photos. Various color spectrums can be used to change the gray into colored pixels. This process enhances both visual and software analysis because various intensities can be identified by various colors. A range of rainbow colors has been selected for the upper image. Black to white colors have been replaced by the black to purple colors, respectively. In the lower photo, a spectrum from black to yellow has been used. Pore spaces are still black but anhydrite nodules have been changed to yellow color. The background of the image is red with various intensities. This is especially useful when two objects should be recognized, such as this case.

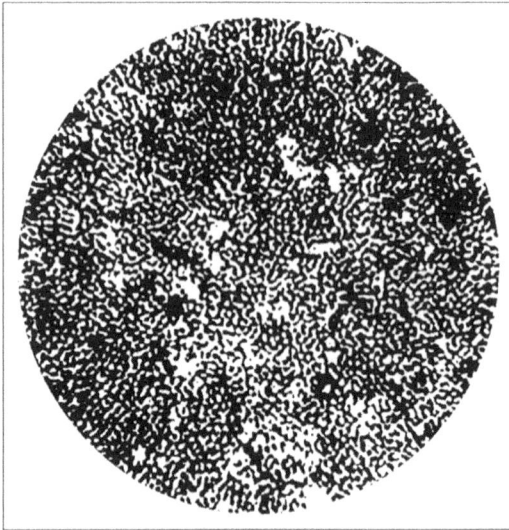

Fig 4.7 Binarization

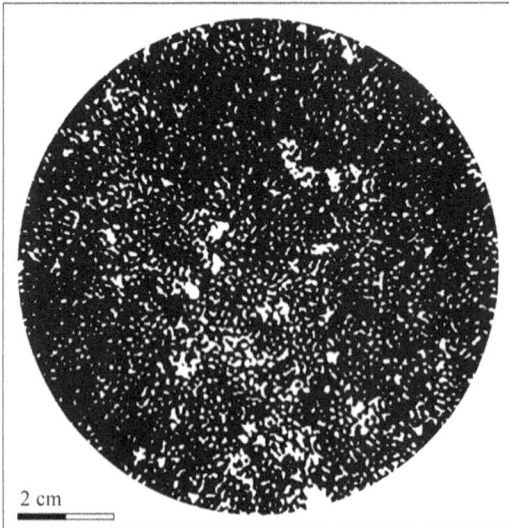

Age: Neocomian
Main lithology: dolomite
Porosity: 22.60 %

FIGURE 4.7
Binarization process includes conversion of all gray-level intensities into two colors, namely black and white. A threshold is defined which separates these two colors. The important point is selecting the value of this threshold. In maximum entropy threshold method, sum of the entropy of the two ranges is maximized. These two ranges are above and below the threshold, respectively. Result is seen in the upper image. In minimum error method (lower image), data distribution is modeled as the sum of two Gaussian distributions. The misclassification error is calculated and the minimum value is selected as the threshold.

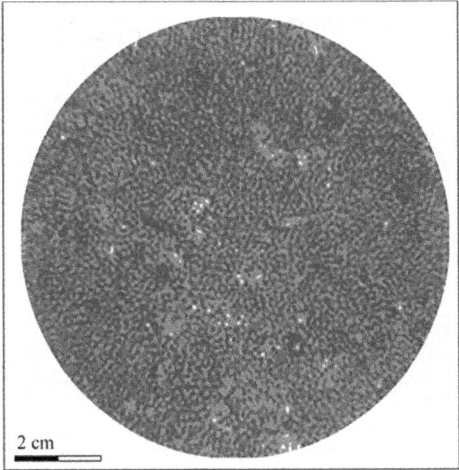

Fig 4.8 K-means clustering

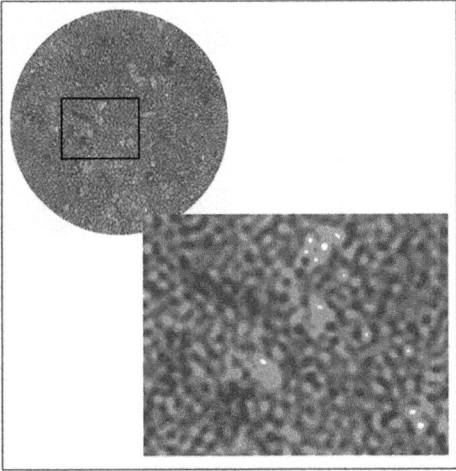

Age: Neocomian
Main lithology: dolomite
Porosity: 22.60 %

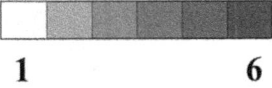

FIGURE 4.8
Clustering is a routine statistical procedure for data grouping. In this photo, k-means clustering has been applied to a CT image and six clusters have been defined. In fact, all pixel values from 0 to 255 have been classified into six groups. This strongly affects the visual and software analysis of the CT slices because various components of the rocks are limited while CT images are composed of a continuous spectrum of gray intensities. Classes 1 and 2 could be assigned to anhydrite nodules and cements while Class 6 is related to the pore spaces. The upper image shows the result of clustering and the lower figure shows a smaller part of the photo with higher zoom.

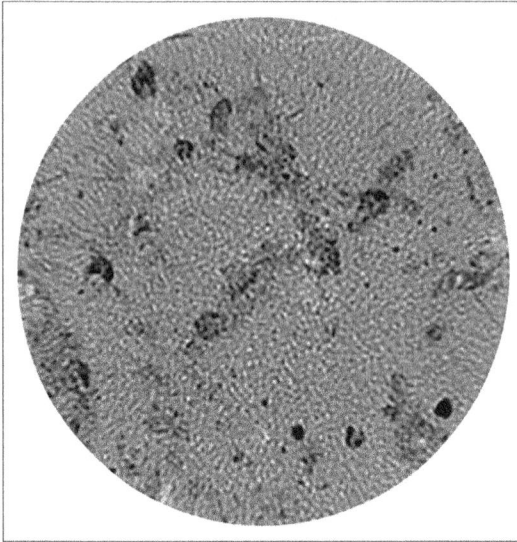

Fig 4.9 Vectorizing

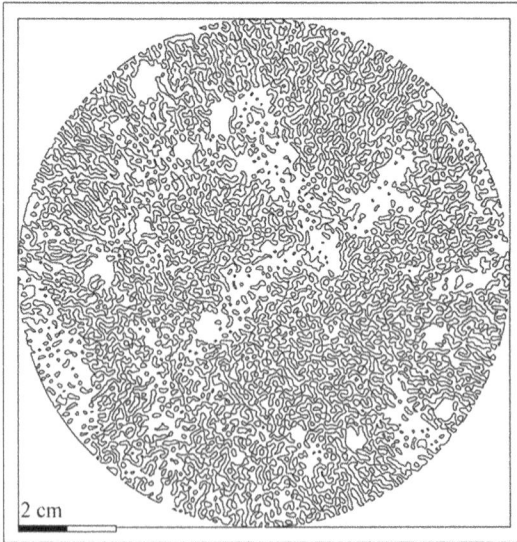

Age: Aptian
Main lithology: dolomite
Porosity: 22.68 %

2 cm

FIGURE 4.9
A CT image is made from pixels with various gray intensities. Therefore, it is a bitmap photo. These photos can be converted to vector graphics. In fact, the image is traced using a software. A threshold is selected and pixels with gray values higher than that value are converted into vector lines. Here, the value of 128 has been selected and the outlines of the objects have been extracted. This method can be used for more accurate analysis of some petrophysical properties, such as calculating the permeability of the sample.

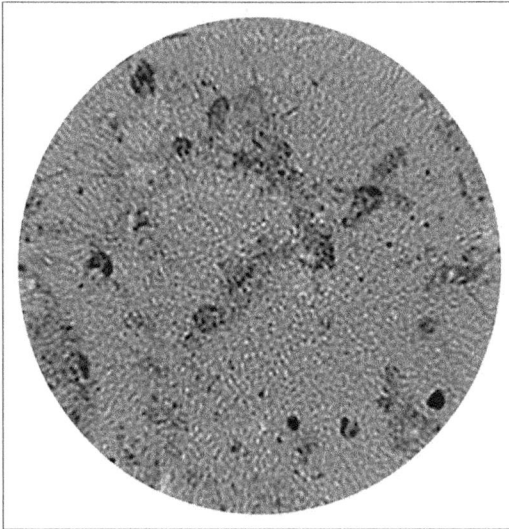

Fig 4.10 Vectorizing

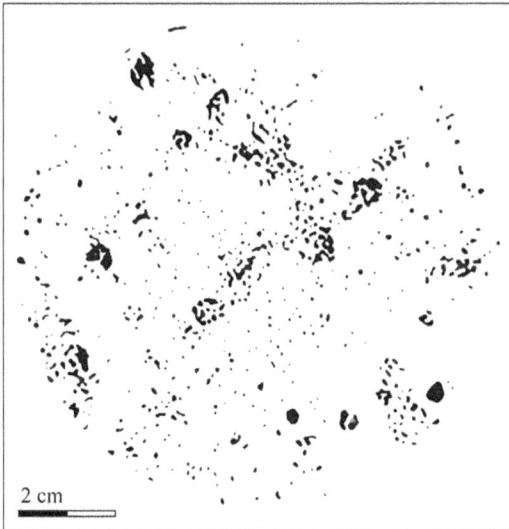

Age: Aptian
Main lithology: dolomite
Porosity: 22.68 %

FIGURE 4.10
The image has been traced with threshold of 60 (gray intensity). As can be seen, the white and even some light gray parts of the image have been removed and empty spaces have been remained. This is very useful for measuring the area of a special feature in a sample, such as porosity. The area of the remained part depends on the selected threshold which is 16.65% in this case. The laboratory measured porosity is 22.68%.

Fig 4.11 Vectorizing

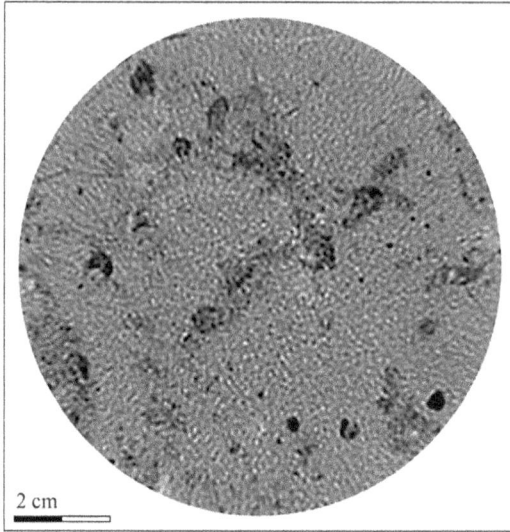

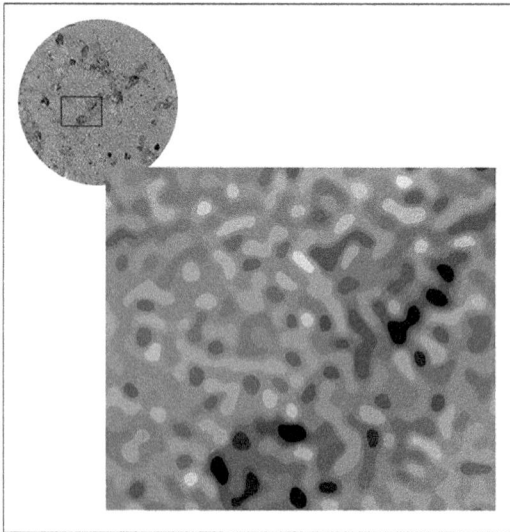

Age: Aptian
Main lithology: dolomite
Porosity: 22.68 %

FIGURE 4.11

The threshold of 50 (gray intensity) has been selected for vectorizing this image with low-fidelity algorithm. This algorithm tries to remove the details of the photo and integrates the pixels with similar gray-color values. The result is seen in the lower image. The area of black rectangle has been zoomed, which shows the result of the used algorithm. Such procedure can be useful for removing the edge effect in CT photos. In fact, the algorithm tries to group all pixels into limited colors (pixels with similar gray intensity values).

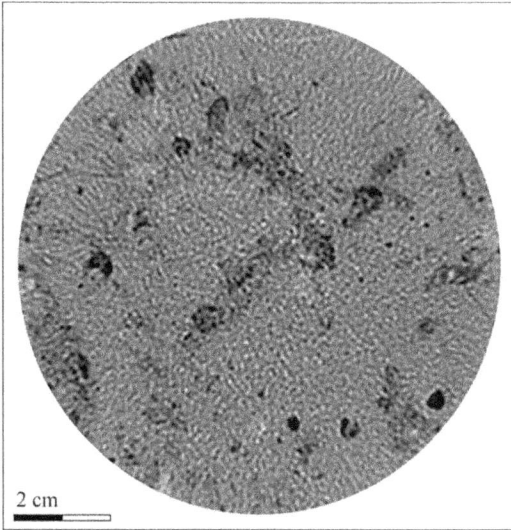

Fig 4.12 Vectorizing

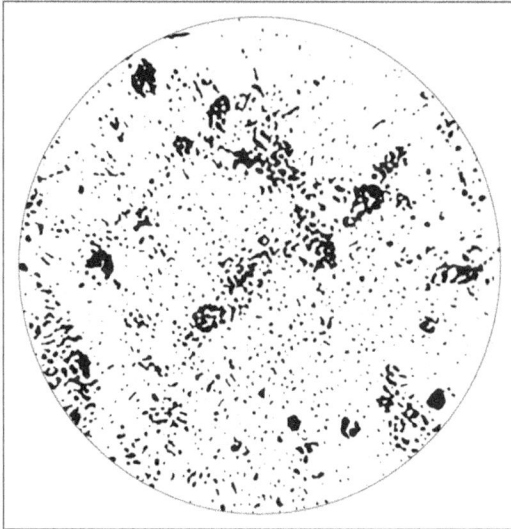

Age: Aptian
Main lithology: dolomite
Porosity: 22.68 %

FIGURE 4.12
With 80 gray intensity threshold, the porosities of the image are clearly seen. Main pore types are moldic and vuggy, but intercrystalline pores are also present. Using this threshold, the black remained area is 23.16%, which is very close to the laboratory measured porosity of the sample (22.68%). The shapes of the pores are also more obvious. This technique is especially useful for calculating microporosities of the samples. Connected dark pixels can be removed with proper methods and single dark spots are considered as microporosities.

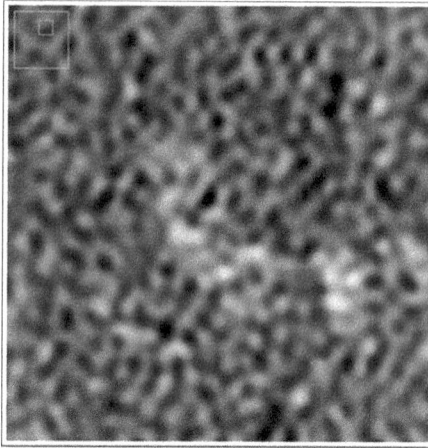

Fig 4.13 Filters (Gussian blur)

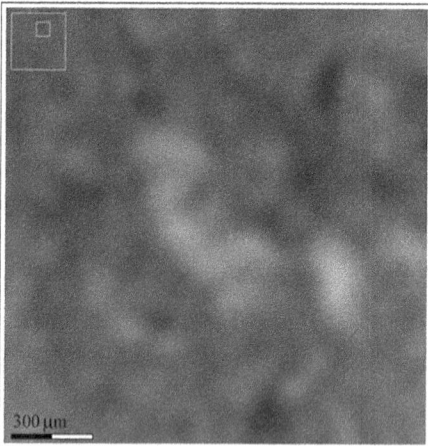

Age: Aptian
Main lithology: dolomite
Porosity: 22.68 %

FIGURE 4.13

Many filters are used to enhance the CT images for further analysis. Gaussian blur filtering is a way to remove noises or details of an image using a Gaussian function. In fact, it uses a Gaussian distribution and eliminates any data out of this normal distribution. In this figure, a small part of a CT image has been selected for this analysis. The original image can be seen in the upper part and the output of this filtering is seen in the lower image. Many details have been eliminated, while some objects are more homogenous.

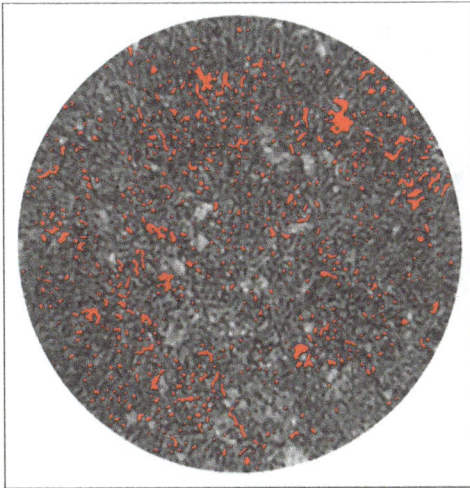

Fig 4.14 Filters (Median)

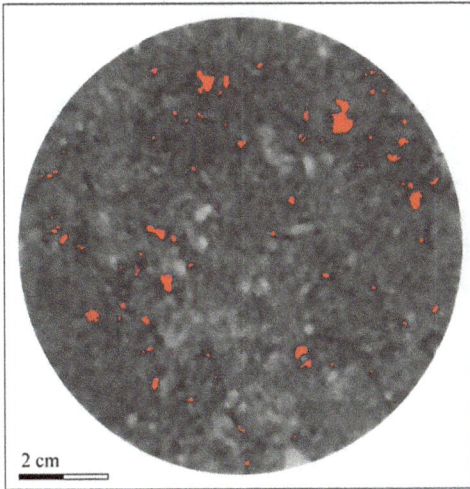

Age: Neocomian
Main lithology: dolomite
Porosity: 22.60 %

FIGURE 4.14
In median filtering, a data window is selected and the values within this window are sorted from small to large. The median of the input values of each window is selected for illustrating the output. The median filter with the two window sizes (5 pixels for the upper image and 10 pixels for the lower one) has been applied to one CT image. Then, pixel values between 0 and 80 have been marked with red color. Results show that increasing the window size decreases the difference in the intensities and so smaller area has been selected.

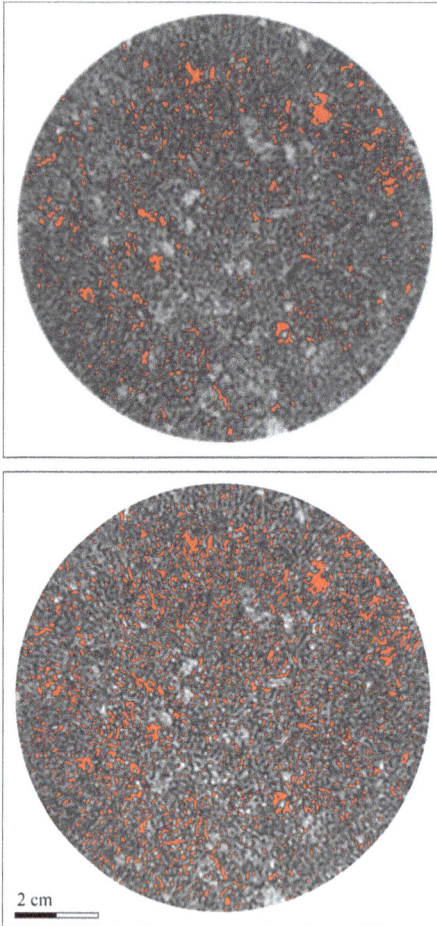

Fig 4.15 Filters (median and Gaussian blur)

Age: Neocomian
Main lithology: dolomite
Porosity: 22.60 %

2 cm

FIGURE 4.15

In image processing, sometimes blurring is needed to reduce the noise or details of an image. This is useful for CT scan images, when the boundary pixels must be added to the object or the matrix. Anyway, the resulted image has objects with sharp boundaries. In the upper photo, a Gaussian filter has been applied to the image and then pixels between 0 and 80 have been colored with red. The lower photo shows the same image with the same threshold, but the median filter has been used with 3 pixels window. It is clear that less area has been selected, because the boundary pixels of the objects have been removed.

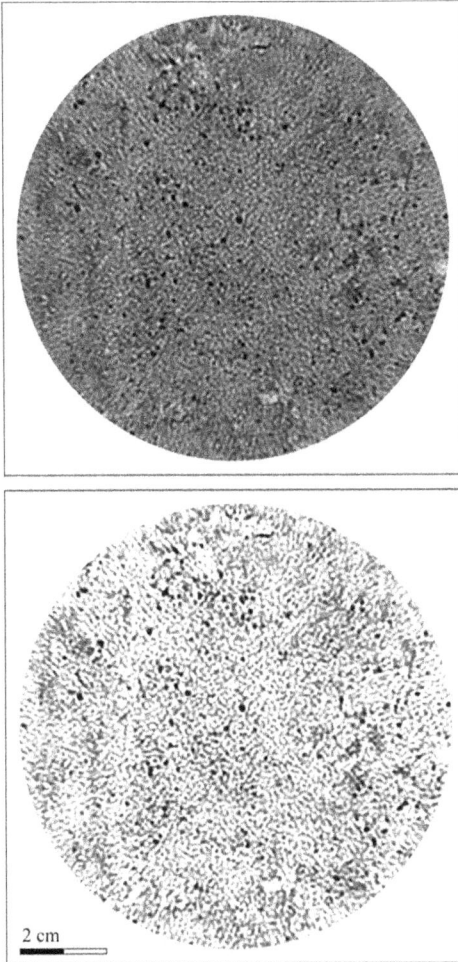

Fig 4.16 Calculation (add)

Age: Oligocene–Miocene
Main lithology: dolomite
Porosity: 21.17 %

2 cm

FIGURE 4.16
Pixels have values in every image. In a full color photo, each pixel has three values for the primary colors (red, green and blue), but in a CT scan image, each pixel has one value between 0 (black) and 255 (white). Mathematical calculations can be applied to these values. These calculations may enhance or reduce the effect of pixel values. For example, the values of this CT image have been added to themselves. In other words, the image has been added to itself. The calculations can be applied to two or more images, but this is meaningless in many cases. In this case, the black parts (near 0 values) are unchanged but the light parts became lighter.

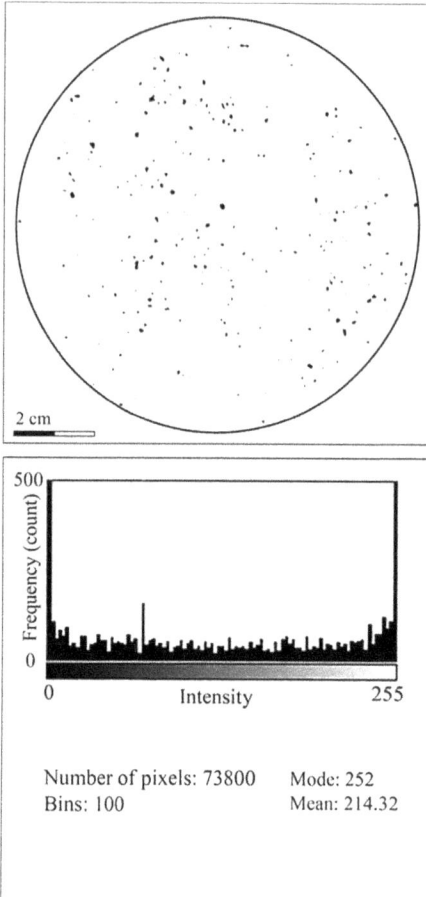

Fig 4.17 Calculation (multiply)

Number of pixels: 73800   Mode: 252
Bins: 100   Mean: 214.32

Age: Oligocene–Miocene
Main lithology: dolomite
Porosity: 21.17 %

FIGURE 4.17
The CT image of the previous page has been multiplied by itself (upper image). This strongly enhances the values. The black parts are still black, while pixels with higher values are white. In this example, the lowest values have remained in the resulted image. The histogram shows the distribution of pixels' frequency in this result. As can be seen, 0 and 255 have the highest values while other frequencies are considerably lower than these two ends. In a normal CT photo, the frequency of pixel values shows a normal distribution (see Figure 4.1). Mode is 252, which is very close to the white color.

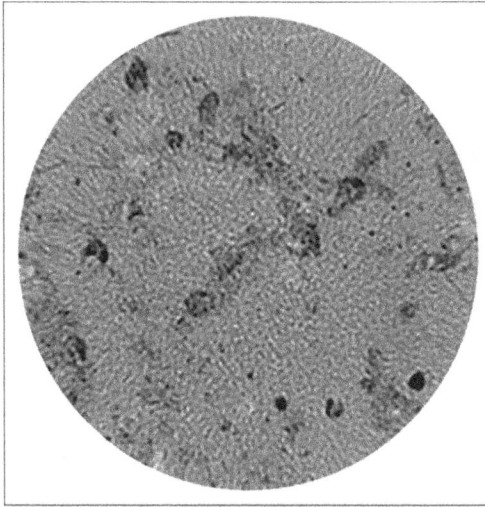

Fig 4.18 Edge detection

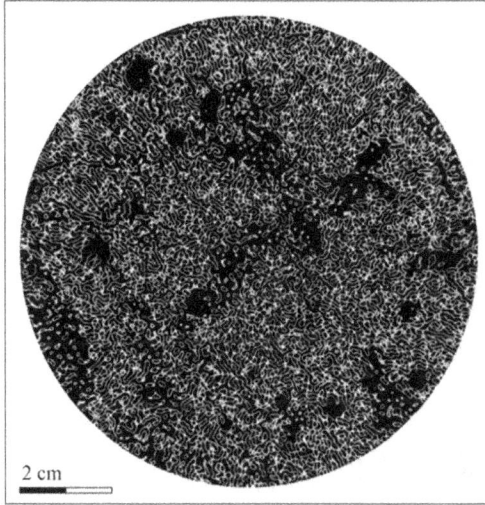

Age: Aptian
Main lithology: dolomite
Porosity: 22.68 %

2 cm

FIGURE 4.18
Edges are sharp changes in image brightness. Therefore, in a CT image, pixels with similar gray values are grouped together. The boundary of these groups is depicted. The upper image shows the original CT photo. An edge detection filter has been applied to the lower photo. The areas of empty spaces are grouped and the boundaries of the objects have been marked with white lines. Such analysis is useful to calculate the connectivity of pore spaces or any other object with enough contrast. Therefore, permeability or pore throat sizes can be determined with this method.

Fig 4.19 Metal noise

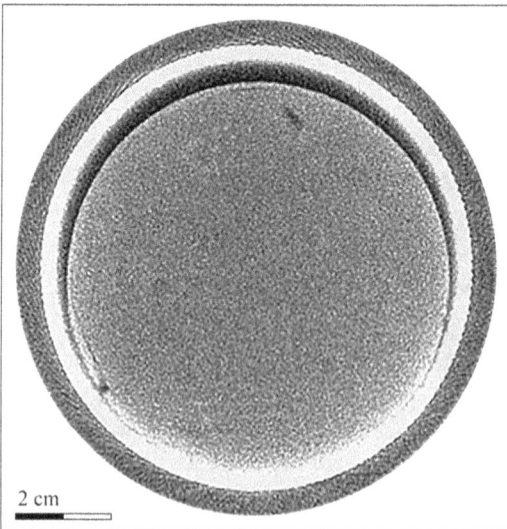

Age: Neocomian
Main lithology: lime
Porosity: 12.81 %

2 cm

FIGURE 4.19
Metal noises destruct the image structure and texture in CT analysis. In the upper image, a metal fastener has been used. It is obviously seen that the distortion has started from the upper right corner of the image, which is the location of the fastener, and attenuates towards the lower left corner. These images are not usable in CT image analysis. The top and bottom of each meter of core are capped and then fastened by such fasteners. Therefore, such image noises are routine in the beginning and end of core CT scan of each meter. In the lower image, a metal tube has been used, which connects two parts of core sleeves. The noise is more homogenous in this image.

Index

Taylor & Francis Group
an **informa** business

Taylor & Francis eBooks

www.taylorfrancis.com

A single destination for eBooks from Taylor & Francis
with increased functionality and an improved user
experience to meet the needs of our customers.

90,000+ eBooks of award-winning academic content in
Humanities, Social Science, Science, Technology, Engineering,
and Medical written by a global network of editors and authors.

TAYLOR & FRANCIS EBOOKS OFFERS:

A streamlined
experience for
our library
customers

A single point
of discovery
for all of our
eBook content

Improved
search and
discovery of
content at both
book and
chapter level

REQUEST A FREE TRIAL
support@taylorfrancis.com

Routledge
Taylor & Francis Group

CRC Press
Taylor & Francis Group

For Product Safety Concerns and Information please contact our EU
representative GPSR@taylorandfrancis.com
Taylor & Francis Verlag GmbH, Kaufingerstraße 24, 80331 München, Germany

www.ingramcontent.com/pod-product-compliance
Lightning Source LLC
Chambersburg PA
CBHW070722220326
41598CB00024BA/3266